The Hand that Feeds

The Hand that Feeds

The complex relations of human–animal feeding

Edited by Alexander Mullan, Riley Smallman, Herre de Bondt and Juliette Waterman

First published in 2025 by
UCL Press
University College London
Gower Street
London WC1E 6BT

Available to download free: www.uclpress.co.uk

Collection © Editors, 2025
Text © Contributors, 2025

The authors have asserted their rights under the Copyright, Designs and Patents Act 1988 to be identified as the authors of this work.

A CIP catalogue record for this book is available from The British Library.

Any third-party material in this book is not covered by the book's Creative Commons licence. Details of the copyright ownership and permitted use of third-party material is given in the image (or extract) credit lines. Every effort has been made to identify and contact copyright holders and any omission or error will be corrected if notification is made to the publisher. If you would like to reuse any third-party material not covered by the book's Creative Commons licence, you will need to obtain permission directly from the copyright owner.

This book is published under a Creative Commons Attribution-NonCommercial 4.0 International licence (CC BY-NC 4.0), https://creativecommons.org/licenses/by-nc/4.0/. This licence allows you to share and adapt the work for non-commercial use providing attribution is made to the author and publisher (but not in any way that suggests that they endorse you or your use of the work) and any changes are indicated. Attribution should include the following information:

Mullan, A., Smallman, R., De Bondt, H. and Waterman, J. (eds.). 2025. *The Hand that Feeds: The complex relations of human–animal feeding*. London: UCL Press. https://doi.org/10.14324/111.9781800088337

Further details about Creative Commons licences are available at https://creativecommons.org/licenses/

ISBN: 978-1-80008-831-3 (Hbk)
ISBN: 978-1-80008-832-0 (Pbk)
ISBN: 978-1-80008-833-7 (PDF)
ISBN: 978-1-80008-834-4 (epub)
DOI: https://doi.org/10.14324/111.9781800088337

This book is dedicated to the whole 'From "Feed the Birds" to "Do Not Feed the Animals"' research team, all our collaborators, interlocutors, supervisors and supporters – and of course, to all the animals we have fed along the way.

Contents

List of figures ix
List of tables x
List of contributors xi
Acknowledgements xii

Need to feed: an introduction to underrepresented aspects of animal feeding 1
Gaia Mortier and Felix Sadebeck

1. A raw egg on an empty stomach: feeding ill cattle in Roman antiquity 9
   Felix Sadebeck

2. Food for flight: feeding of captive raptors in medieval Britain 25
   Hannah Britton and Arthur Redmonds

3. 'I live off them, they live off me': exploring the human–flea feeding relationship in the history of flea circuses 45
   Gaia Mortier

4. Human–raptor relationships in urban spaces: the history of red kites (*Milvus milvus*) and human food in Britain 61
   Juliette Waterman

5. Feed the birds but stone the crows: the role of food in conflict with corvids throughout British history 81
   Riley Smallman

6. Whose food, whose health? Moral and ecological hierarchies of urban stray cats and pigeons 105
   Giovanna Capponi and Herre de Bondt

7. Feeding farm animals: perceptions and performances of the 'good farmer' among regenerative farmers 123
   Hannah C. Mortimer

8. The adventures of a birch branch; or, a narrative ethnography of browse feeding at the Highland Wildlife Park 147
   Alexander Mullan

9 You are what you eat: dietary drivers of morphological change 161
David Cooper and Andrew C. Kitchener

10 The effects of red fox chronic exposure to metals on health and the environment 189
Blessing Chidimuro

11 The pros, cons and contrary consequences of conservation feeding: anthropogenic feeding of the red kite (*Milvus milvus*) in Britain 205
Virginia Thomas

Conclusion. In conversation: non-utilitarian feeding, interdisciplinarity and the future of feeding research 219
Herre de Bondt, Hannah Britton, David Cooper, Gaia Mortier, Hannah C. Mortimer, Felix Sadebeck, Virginia Thomas and Juliette Waterman

Index 237

List of figures

4.1	Detail from the illuminated folio 46v of the *Aberdeen Bestiary* MS. 24, showing a red kite.	66
5.1a	Abundance of British archaeological corvid remains, by period.	89
5.1b	Abundance of British artefacts depicting corvids, by period.	90
5.2	Roman intaglio depicting raven among symbols of Apollo and agricultural prosperity.	91
6.1	A stray cat being fed at the colony of the Magic Door, Piazza Vittorio Emanuele II, Rome.	111
6.2	Sign in Waterlow Park posted by Camden Council (left). Same sign after alterations by SFS member (right).	117
7.1	The Cattle Farmer walks through his herd of Hereford cattle to set up a new paddock for them to graze.	141
9.1	Diagram of skull and musculature. Line drawings were generated from Brutus the lion, a male that was imported from South Africa as an adult and lived at Edinburgh Zoo until he died in March 1923.	171
9.2	A tiger pulling meat from a feeding pole at Glasgow Zoo.	174
10.1	Boxplots showing the comparisons of heavy metal concentrations – lead (Pb), cadmium (Cd), arsenic (As) and chromium (Cr) – in red foxes in London between the 1970s and the 2020s.	196

List of tables

1.1	Ingredients listed by Cato (*De agri cultura*, chapters 70–3) as part of recipes for healing cattle.	15
4.1	Animals listed in the Tudor vermin law of 1566 (Eliz an.8, cap.XV), in order of appearance and with stated bounty.	68
5.1	Summary of earliest corvid persecution and accusations by species.	95
9.1	Effects of nutrition upon skeletal development, health and morphology.	167
9.2	Experiments designed around the effects of manipulation of the physical properties of food on the mechanical response of the skull and mandible, to determine how these structures may vary plastically due to variation in stresses.	170
9.3	The mechanical properties of food, following definitions from Berthaume 2016.	172
9.4	The typical behaviours associated with the hunting, killing and eating of prey by wild tigers compared with behaviours elicited by different kinds of captive diet and the tiger feeding pole.	176
11.1	Types of wildlife feeding.	206

List of contributors

Alexander Mullan is a PhD student at the University of Roehampton, investigating the dynamics of human–animal feeding within zoos.

Riley Smallman is a zooarchaeological PhD student at the University of Exeter, exploring prehistoric to modern-day British human–avian relationships.

Herre de Bondt is a PhD student at the University of Roehampton, exploring urban human–animal relationships through an anthropological investigation of bird feeding.

Juliette Waterman is a PhD student in archaeology at the University of Reading, analysing the diets of reintroduced birds in the UK through time.

Hannah Britton is a zooarchaeology PhD student at the University of Exeter with a focus on the deep-time history of falconry birds.

Giovanna Capponi is an assistant professor at the State University of Rio de Janeiro; her PhD and postdoctoral research focused on environmental anthropology and more-than-human ontologies.

Blessing Chidimuro conducted her postdoctoral research at the University of Reading, using isotopic analyses to explore animal diets.

David Cooper is a postdoctoral researcher at National Museums Scotland whose work focuses on morphological variation in vertebrates due to diet.

Andrew C. Kitchener is Principal Curator of Vertebrates at National Museums Scotland, a member of the Scottish Animal Welfare Commission, and is interested in the effects of captivity on the morphology of mammals and birds.

Gaia Mortier is a PhD student at the University of Reading, investigating biochemical signals in ectoparasites found in museum and archaeological collections.

Hannah C. Mortimer is a PhD student at the University of Exeter, using ethnography to explore different aspects of care on regenerative farms.

Arthur Redmonds is a medieval archaeology PhD student at the University of Exeter researching castle landscapes.

Felix Sadebeck is a zooarchaeology PhD student at the University of Exeter studying the morphologic diversity of Roman cattle in southern England.

Virginia Thomas works at AgResearch, Aotearoa New Zealand; her postdoctoral research investigated the conservation feeding of animals in relation to reintroduction programmes.

Acknowledgements

This work was supported by the Wellcome Trust (grant number 219889/Z/19/Z) and the Arts and Humanities Research Council via South, West and Wales Doctoral Training Partnership (in the cases of Hannah Britton, Arthur Redmonds, Felix Sadebeck, and Riley Smallman). Research was conducted within the institutions and facilities of the University of Exeter, the University of Reading, the University of Roehampton and National Museums Scotland.

We thank Rosanne Strachan Law for permission to reproduce the photograph by the late Graham Law of a tiger on a feeding pole (Figure 9.2).

Need to feed: an introduction to underrepresented aspects of animal feeding

Gaia Mortier and Felix Sadebeck

Whether it be tossing slightly stale chunks of bread in a pond for the local ducks and geese, hand-feeding goats hay at the petting zoo, or offering urban rats a feast by accidentally dropping crumbs from your croissant on the go, feeding truly is the basis of many human–animal interactions. Animal feeding is a global phenomenon which every person almost certainly participates in, whether intentionally or not, driven by a range of different motivations and experienced in a multitude of contexts. Providing food to animals brings us closer to them, both physically and emotionally, and creates opportunities to alter their – and our – behaviour, be it for better or worse. It has consequences for the environment and health of the animals we feed, those we do not feed, and ourselves. The ways in which we go about it – and the animals we deem worthy of our attention – are deeply interconnected with our world views, and give insight into our evaluation of other species and 'human' food resources.

Despite this, previous animal feeding studies have overwhelmingly focused on utilitarian and effective aims – that is, feeding for the intentional generation of animal-derived products such as meat, eggs, milk, leather and traction (for instance, see Saha and Pathak 2021; Kumar et al. 2014; Kiczorowska et al. 2017; Theodossopoulos 2005) – with additional aspects mostly concerning pet feeding (see Irvine and Cilia 2017; Michel et al. 2008; Dodd et al. 2020) and the human-health implications of animal feeding (such as in Finley et al. 2006; Sapkota et al. 2007; Huss et al. 2018). Non-utilitarian and affective – that is, emotional and non-product-driven – aspects of human–animal interactions have been investigated far less commonly (examples include Oma 2010; Hill 2011; Sykes 2014, especially chapter 7), particularly where feeding is concerned. When elements of affective feeding have been discussed, the focus is on the wider relationship rather than the act of feeding specifically. This book explores these issues but will not try to answer a simplified question of whether it's 'good' or 'bad' to feed animals – we aim to open discussions, instead of closing them.

Although the emotional motivations and effects of feeding have previously been considered in regard to pets (for instance in Martinez-Caja et al. 2022; Delime et al. 2020; Tsai et al. 2020), they are also primary drivers behind feeding other species. Furthermore, these affective motivations underlie many other superficially utilitarian feeding practices of non-pet domestic species, as will be explored in this book. To do so, it is vital to acknowledge that animals are subjective agentic beings (Thomas 2016): how they behave, how they allow us to interact with them and crucially whether they choose to eat is largely within their control in many encounters. However, feeding presents a crucial juncture wherein humans assert control over animals, for example by restricting which foods are 'appropriate' for certain species and groups.

It is important to acknowledge that the labels we give to certain animals – such as wild, domestic, feral, tame and pest – are both anthropocentric and contentious. While we understand wild species to be those not owned or managed by humans, living freely and unconstrained, anthropogenic land usage and species population control efforts do inevitably restrict such animals and bring them closer to people, consequently developing commensal and/or synanthropic relationships with us and our human food sources. In these contexts, a clear distinction between wild and domestic seems difficult: if an animal born into the wild becomes adjusted and accustomed to humans enough to be considered tame in a behavioural sense, is it still wildlife? If a domestic animal escapes or is abandoned, we term it feral – a word positioned somewhere between domestic and wild, but imbued with negative connotations. This is especially the case where feral and pest border each other: feral or wild animals may be designated as pests, worthy of disdain, exclusion or outright extermination, based on a plethora of reasons that stray far beyond direct human–animal interactions into historical, cultural and political influences. The aim here is not to disentangle these complicated, interwoven terms, but rather to explore how acts of feeding ultimately determine, reinforce and challenge them.

By looking through the lens of feeding, we can see how categories like affective or effective, utilitarian or non-utilitarian, and even domestic or wild, enforce binary ways of thinking that oversimplify complex interactions and limit our ability to explore them holistically. Analysing these complexities opens new interpretational pathways for understanding human–animal interactions and incites discussion of previously overlooked but promising and fascinating aspects of animal feeding.

This book results from the Wellcome-funded Animal Feeding project 'From "Feed the Birds" to "Do Not Feed the Animals"'. The project set out to undertake a deep-time and cross-cultural investigation to uncover the roots of animal feeding and critique the benefits and risks for all concerned. It was born from the hypothesis that animal domestication itself was driven by the human instinct to feed animals, and that this process is not just continuing but accelerating, with consequences for human, animal and environmental health. For example, while we are not intentionally domesticating wild birds, the increasingly popular practice of bird feeding results in consequences for us, the birds and the larger ecosystems we live in (Doremus et al. 2023). It could be argued that garden birds are not wild any more as a result of feeding (Schilthuizen 2018), but describing them as domestic or even tame does not seem quite appropriate either. As we increasingly scrutinise human impacts on ecologies and our place within the natural world, now is the critical time to evaluate the intricacies of animal feeding and its vital role in human–animal interactions.

The aim to undertake a comprehensive investigation of predominantly non-utilitarian animal feeding practices required an interdisciplinary team of researchers, as the multiple facets of animal feeding are often found at the intersections between various subjects. Specifically, the fields of zoology, bioarchaeology and anthropology formed the basis for the project, incorporating an array of subjects and methodologies. Each collaborator brought with them their own perspective and expertise, which, combined with the availability of material provided by the project partners, resulted in the investigation of a range of animal species. It must be emphasised that the analysed species are not meant to represent an exhaustive list, but to serve as illustrative examples that demonstrate how new discussions can be opened through investigating the affective aspects of animal feeding with interdisciplinary perspectives. The locations of our researchers additionally limit the scope to covering the Global North, and in particular to inspecting the interactions between people and animals through food within north-western European cultures. The diversity of species and situations in which animals are fed varies enormously around the world, and the scope of the project highlighted just a few of them, with many more opportunities for further investigation. The authors of this book thoroughly encourage other researchers and animal-feeders from around the globe to contribute their perspectives and case studies through future work to continue the discussions we begin here.

While unique in its approach to exploring non-utilitarian feeding from a wide range of interdisciplinary perspectives, this book is situated within broader scholarly literature of the vast field of human–animal studies. Some notable works that have also recognised the crucial role feeding plays in interspecies interactions include Terry O'Connor's *Animals as Neighbors: The past and present of commensal species* (2013), which integrates anthropological exploration with archaeological excavation, telling stories of commensalism across multiple species, various regions and through time, while questioning the terminologies and categories used to define such relationships. With a utilitarian focus but critical to understanding themes of the broader impacts of feeding, Marion Nestle's *Pet Food Politics: The chihuahua in the coal mine* (2008) exposes the entanglement of food production systems intended for pets, farm animals and people alike, revealing the complexities and concerns surrounding the allocation of global food resources and in particular food-safety management. Meanwhile, Josh Milburn's *Food, Justice, and Animals: Feeding the world respectfully* (2023) questions whether the development of ethical farming systems could lead to a non-vegan future that is still respectful of animal rights and ecologically sustainable, further integrating animal agency into discussions surrounding food. Within these same realms, Elan Abrell's *Saving Animals: Multispecies ecologies of rescue and care* (2021) ethnographically examines animal sanctuaries and their role in our understandings of ethical treatments of animals, questioning what 'care' and 'sanctuary' really mean within modern human–animal ecologies. While these sources cover systemic, ethical, and moral facets of human–animal relations, our book instead zooms in on the intimate practice of non-utilitarian feeding. The book presents a wide array of novel multispecies, multi-context, multidisciplinary studies that use feeding to build upon prior literature from new perspectives, generating new lines of enquiry in the process.

In conjunction, the chapters weave a red thread through the maze of topics explored within the Wellcome-funded project. They discuss the underlying themes, tackling the tangled terminology and presenting case studies that explore many of the newly discovered interpretational pathways through which we feed animals. A multidisciplinary foray into the human activity of providing or denying food to other species is presented, achieved by considering the perspectives and agency of animals and reflecting on non-utilitarian, affective motivation in human actions. This book seeks to re-evaluate the biases and assumptions that surround the feeding of animals, without offering simple solutions but rather demonstrating the richness and progressive power of approaching the topic from new viewpoints, instigating new discussions in human–animal studies. As such, the chapters do not offer final interpretations of specific human–animal interactions but rather explore new pathways that can be trodden by future research.

Although perhaps just a starting point, these pathways are taken up by our authors, offering various perspectives in the following chapters. Illustrating the overlap and contrast between effective and affective feeding relationships, in Chapter 1 Felix Sadebeck analyses Roman veterinary feeding of cattle and how such interactions transgress human–animal divides through food. Chapter 2, by Hannah Britton and Arthur Redmonds, considers medieval falconry relationships, epitomised through the unusual and exotic cures fed to sick birds. Continuing on with peculiar feeding practices, Gaia Mortier tours the history of flea circuses in Chapter 3, where parasitic pests were repurposed as entertainment by feeding upon their owners' blood. Juliette Waterman examines the shift in public perceptions of red kites from pests to iconic reintroduced species in Chapter 4, demonstrating the vital links between food scavenging behaviours and how we value a species. Riley Smallman continues this line of exploration in Chapter 5 with a case study of how food is a source of conflict with corvids, and Chapter 6 further investigates which species are deemed worthy of feeding through anthropological studies by Giovanna Capponi and Herre de Bondt on stray cats and pigeons. Chapter 7 demonstrates the difficulty in separating utilitarian from non-utilitarian feeding as Hannah Mortimer explores modern farming practices from an anthropological perspective. Both Chapters 8 and 9 focus on captive animals, where Alexander Mullan observes the anthropological intricacies of zoo feeding and David Cooper and Andrew Kitchener examine morphological differences between zoo-fed animals and their wild counterparts. In Chapter 10 Blessing Chidimuro reveals the impacts of urban food scavenging by analysing metallic pollutants within fox bones. Virginia Thomas explores the effects of human-driven conservation feeding on species on the brink of extinction with a case study of red kites in Chapter 11. Finally, the book's conclusion – an 'in conversation' piece between the authors – reflects on the diversity of feeding practices, the nature of interdisciplinary research and future avenues of study. Through the variety of applied methods, different viewpoints and case studies, the chapters of this book synergise into a kaleidoscope, offering new perspectives on previously neglected aspects of animal feeding.

References

Abrell, Elan. 2021. *Saving Animals: Multispecies ecologies of rescue and care*. Minneapolis: University of Minnesota Press.

Delime, Perrine, Kadri Koppel, Pascal Pachot and Aurelie De Ratuld. 2020. 'How the odor of pet food influences pet owners' emotions: A cross cultural study', *Food Quality and Preference* 79 (103772): 1–10.

Dodd, Sarah, Nick Cave, Sarah Abood, Anna-Kate Shoveller, Jennifer Adolphe and Adronie Verbrugghe. 2020. 'An observational study of pet feeding practices and how these have changed between 2008 and 2018', *Veterinary Record* 186 (19): 623–56.

Doremus, Jacqueline, Liqing Li and Darryl Jones. 2023. 'Covid-related surge in global wild bird feeding: Implications for biodiversity and human–nature interaction', *PLOS ONE* 18 (8, e0287116): 1–15.

Finley, Rita, Richard Reid-Smith and J. Scott Weese. 2006. 'Human health implications of Salmonella-contaminated natural pet treats and raw pet food', *Clinical Infectious Diseases* 42 (5): 686–91.

Hill, Erica. 2011. 'Animals as agents: Hunting ritual and relational ontologies in prehistoric Alaska and Chukotka', *Cambridge Archaeological Journal* 21 (3): 407–26.

Huss, Anne, Roger Cochrane, Cassie Jones and Griffiths G. Atungulu. 2018. 'Physical and chemical methods for the reduction of biological hazards in animal feeds'. In *Food and Feed Safety Systems and Analysis*, edited by Steven C. Ricke, Griffiths G. Atungulu, Chase E. Rainwater and Si Hong Park, 83–95. Cambridge, MA: Academic Press.

Irvine, Leslie and Laurent Cilia. 2017. 'More-than-human families: Pets, people, and practices in multispecies households', *Sociology Compass* 11 (e12455): 1–13.

Kiczorowska, Bożena, Wioletta Samolińska, Ali Ridha Mustafa Al-Yasiry, Piotr Kiczorowski and Anna Winiarska-Mieczan. 2017. 'The natural feed additives as immunostimulants in monogastric animal nutrition: A review', *Annals of Animal Science* 17 (3): 605–25.

Kumar, Muneendra, Vinod Kumar, Debashis Roy, Raju Kushwaha and Shalini Vaiswani. 2014. 'Application of herbal feed additives in animal nutrition: A review', *International Journal of Livestock Research* 4 (9): 1–8.

Martinez-Caja, Ana Martos, Veerle De Herdt, Marie-Jose Enders-Slegers and Christel Palmyre Henri Moons. 2022. 'Pet ownership, feelings of loneliness, and mood in people affected by the first COVID-19 lockdown', *Journal of Veterinary Behavior* 57: 52–63.

Michel, Kathryn E., Kristina N. Willoughby, Sarah K. Abood, Andrea J. Fascetti, Linda M. Fleeman, Lisa M. Freeman, Dorothy P. Laflamme, Cassondra Bauer, Brona L. E. Kemp and Janine R. Van Doren. 2008. 'Attitudes of pet owners toward pet foods and feeding management of cats and dogs', *Journal of the American Veterinary Medical Association* 233 (11): 1699–703.

Milburn, Josh. 2023. *Food, Justice, and Animals: Feeding the world respectfully*. Oxford: Oxford University Press.

Nestle, Marion. 2008. *Pet Food Politics: The chihuahua in the coal mine*. Berkeley: University of California Press.

O'Connor, Terry. 2013. *Animals as Neighbors: The past and present of commensal animals*. East Lansing: Michigan State University Press.

Oma, Kristin Armstrong. 2010. 'Between trust and domination: Social contracts between humans and animals', *World Archaeology* 42: 175–87.

Saha, Subodh Kumar and Nitya Nand Pathak. 2021. *Fundamentals of Animal Nutrition*. Singapore: Springer.

Sapkota, Amy R., Lisa Y. Lefferts, Shawn McKenzie and Polly Walker. 2007. 'What do we feed to food-production animals? A review of animal feed ingredients and their potential impacts on human health', *Environmental Health Perspectives* 115 (5): 663–70.

Schilthuizen, Menno. 2018. *Darwin Comes to Town: How the urban jungle drives evolution*. London: Quercus.

Sykes, Naomi. 2014. *Beastly Questions: Animal answers to archaeological issues*. London: Bloomsbury.
Theodossopoulos, Dimitrios. 2005. 'Care, order and usefulness: the context of the human–animal relationship in a Greek island community'. In *Animals in Person: Cultural perspectives on human–animal intimacies*, edited by John Knight, 15–35. London: Bloomsbury.
Thomas, Natalie. 2016. 'Animals as agents'. In *Animal Ethics and the Autonomous Animal Self*, by Natalie Thomas, 7–36. London: Palgrave Macmillan.
Tsai, Weilun, Martin Talavera and Kadri Koppel. 2020. 'Generating consumer terminology to describe emotions in pet owners and their pets', *Journal of Sensory Studies* 35 (5): e12598.

1
A raw egg on an empty stomach: feeding ill cattle in Roman antiquity
Felix Sadebeck

Introduction

The distinction between utilitarian and non-utilitarian feeding seems to be clear only at first glance: utilitarian (or effective) feeding includes fattening an animal to produce meat or other products – to produce an effect – whereas non-utilitarian feeding, such as giving treats, may seem to have no direct function aside from the affective and emotional response. However, both aspects are closely linked. To give an example: when Roman herdsmen tried to heal their diseased cattle by feeding them specific food, the original intention can be considered to be utilitarian, but these acts were actually intricately woven with affective intentions and effects as well. In these moments of intense care, the herdsman and the animals engage in intricate and emotional bonds. Treatment becomes ritual: ritual levels the former hierarchy between livestock and herdsman towards a more horizontal relationship, and the emotional impact far exceeds the original utilitarian purposes, transgressing towards a non-utilitarian feeding relationship. These emotional 'side-aspects' of feeding cattle can, in an altered form, still be observed in present-day cattle farms (see Chapter 7).

Previous works concerned with Roman-period veterinary treatises have discussed the actual medical properties of the veterinary recipes (for instance Mezzabotta 2001, 143–8) or focused on philological discussions of the literary traditions behind the fundamental sources (Fischer 1979; Fischer et al. 1998; Adams 1995; Viré 2007; Bertocchi and Orlandini 2007). However, there is previously untapped scope to use this material as a lens to examine exciting aspects of human–cattle relationships.

Roman agricultural writers offer us one of the earliest comprehensive examples for understanding past human–cattle relationships in moments of veterinary care (Mezzabotta 2001, 148–9; Rex 1998, 12). This is because these writers incorporated earlier advice on cattle husbandry, originating in North Africa (Heurgon 1976; Baldwin 1963, 788 fn. 2; Mahaffy 1889) and ancient Greece (Fischer 2020; Diederich 2007, 56). This allows us to take a glance far back into Mediterranean and Eurasian history, utilising the fine details offered by written accounts. This is important for the topic, as it is problematic to make meaningful interpretations about the subtle emotional interactions unfolding in feeding ill cattle based on archaeological evidence alone.

Cato the Elder (234–149 BCE) and Columella (4–c.70 CE) were selected as examples of Roman agricultural writers, as both have had considerable impacts on later agricultural treatises (see Adams 1995, 13, for Cato's influence on the writings of Pelagonius; Adams 1995, 23, for Columella's influence on the writings of Palladius; Rex 1998, 12–127, for an overview). This allows us to treat them as being fairly representative of general Roman trends, making them a reasonable choice for this case study. Furthermore, the nearly three centuries spanned by the works of those two writers allows us to consider changes and continuities within the Roman period, instead of casting a spotlight on only a single moment of this long period.

Utilising the combined potential of Cato's and Columella's works can open new paths for understanding Roman human–animal relationships. Treading these paths reveals the significance of emotional interactions and the non-utilitarian qualities of veterinary treatment, and can challenge how we interpret archaeological evidence or think about modern-day veterinary practices.

Feeding as veterinary treatment

Knowledge of Roman veterinary treatments has been poorly transmitted into our times, due to a lack of interest in veterinary topics during late antiquity and the early medieval period (Fischer 2023, 126). Consequently, Roman veterinary knowledge is a very remote scholarly field to this day (Fischer et al. 1998). Even Vegetius, a fourth-century writer of a veterinary compendium for horses, *Digesta Artis Mulomedicinae* (including a brief chapter on bovine diseases – see Ortoleva 2008), deemed it still necessary to justify the chosen topic of his book by emphasising that one can learn about human medicine by studying the health of horses (Veg. *mulom*. prol. 9 in Ortoleva 1999). Due to this former lack of interest for the specific subject, a large part of our knowledge about everyday veterinary practice in the Roman world derives from books that were of more interest to antique and medieval scribes: agrarian handbooks.

Unsurprisingly, therefore, Roman veterinary practices can be divided into two fields. For serious issues, especially concerning expensive animals like sport or military horses, one might call a *medicus pecorum*, a professional vet who has likely studied the specific treatises on the topic, comparable to Vegetius' compendium on horses (see Varr. 2.1.21 in Goetz 1929; Col. 7.3.16 in Ash 1941). As mentioned above, there were some attempts to evaluate this Roman veterinary medicine in the mirror of modern knowledge but, as Walker (1973) puts it, this is of doubtful value as it is an anachronistic comparison and will not enable us to understand the everyday part played by veterinary knowledge and practitioners in Roman life. That is because most ailments of livestock would have been cared for by your *pastor diligens*, your diligent herdsman, whom you would expect to have read a few of the standard agrarian handbooks and treatises on the topic of caring for ill animals (Varr. 2.10.10 in Goetz 1929). These agrarian handbooks, as mentioned above, received a lot of attention – even in medieval times – and therefore tend to come with a more or less acceptable transmission history (Rex 1998, 128–32). This, together with the observation that most everyday veterinary practice would have been conducted by a herdsman schooled in these books, probably makes them the richest sources available for understanding everyday Roman-period veterinary treatments.

Generally, there is only so much that a diligent herdsman can do to treat animals, as serious chirurgic procedures would require a *medicus pecorum*. Therefore, it is no surprise that the advice for treating ill animals in these agrarian handbooks mostly boils down to edible remedies. Columella and Cato offer lengthy descriptions of various recipes that one ought to administer for various illnesses. It is not within the scope of this chapter to discuss the potential medical value of these recipes; rather, I wish to highlight that not only the kind of food given in these moments of care, but also the way it is given, sheds light on intricate relationships unfolding in these specific feeding contexts.

As just one of many potential case studies that could be used to approach this topic, this chapter focuses on recipes intended for treating cattle. There are several reasons for this. First of all, cattle had an overwhelming importance in the Roman world. These animals are closely linked with Roman identity: the name of Italy itself roughly translates to 'land of the calf' (Dion. Hal. *ant.* 1.35.2 in Jacoby 1885; Varr. 1.5.3 in Goetz 1929), and the mythic foundation of Rome was allegedly conducted with cattle (Dion. Hal. *ant.* 1.88.2 in Jacoby 1885). These are just two of many more examples. Furthermore, they are arguably the most important livestock in the Roman world (see King 1978 and Albarella et al. 2008 for an overview of the economic importance of cattle in the Roman provinces, and Grau-Sologesta et al. 2022 for an emphasis on the importance of cattle in frontier regions). Additionally, herdsmen's relationship with their cattle was likely to be especially profound, as they spent such a long time with most of their cattle, repeating intricate moments like milking or ploughing over and over again with the same individuals (Fudge 2013, 26). Also, most Roman-period cattle were not primarily kept for milk or meat but rather for their traction power (see, for example, King 2001, 218–22; Peters 1998, 276; or Cat. 61.6 in Hooper and Ash 1934; Varr. 2.5.3 in Goetz 1929; Cic. *off.* 2.89 in Miller 1913), emphasising the importance of the previous point for our case study. This impression from written evidence is confirmed by the archaeological record, for example in Britain, where cattle had the longest average lifespan by far during the Roman period compared to any other time in history (Sykes 2014, 170–2). Due to the general economic and ideological importance of cattle, and the comparably long time of acquaintance with each individual, the relationship between Roman herdsmen and their cattle is likely to have been characterised by strong emotional bonds, which have been overlooked by previous, economic-centred studies of Roman cattle husbandry.

With this in mind, passages from the agrarian handbooks of Cato the Elder and Columella are used as examples to analyse the relationship between herdsmen and their cattle during moments of intensive care triggered by illness. As this care is mainly enacted through feeding, some examples of what exactly herdsmen are feeding to their ill cattle, and how exactly they administer it, are considered and interpreted as revealing emotional bonds, strong enough to transgress boundaries between humans and animals. These bonds, it will be argued, might have their origins in past rituals. The original utilitarian purpose of feeding 'medicine' to ill cattle in order to achieve a specific medical effect is revealed as a mere cover for the emotional relations unfolding during these moments of care.

Aspects of veterinary feeding

Let us begin with some examples from Cato the Elder, Rome's earliest agricultural writer whose work has come down to our times (Mezzabotta 2001; Fischer 2020); indeed, Cato's work is the oldest piece of Latin prose existent today (Rex 1998, 12). In his book *De agri cultura*, he incorporated much content from the most famous treatise on cattle husbandry during his time, a handbook written by the Carthaginian Mago, whose work is unfortunately lost to us but was considered extremely influential for the development of Roman agricultural knowledge beyond oral tradition (Varr. 2.5.18 in Goetz 1929; Reay 2005, 355). Therefore, Cato's recipes for treating ill cattle through feeding can be considered as reflecting a very early stage of the Mediterranean literary tradition concerning the interaction with cattle: trends and tendencies visible in Cato's writings might give us a glance into human–animal relationships of even earlier periods.

In chapters 70–3 Cato gives a few allegedly well-tried recipes for creating a potent remedy for your ill cattle by mixing different kinds of food:

> If you have reason to fear sickness, give the oxen before they get sick the following remedy: 3 grains of salt, 3 laurel leaves, 3 leek leaves, 3 spikes of leek, 3 of garlic, 3 grains of frankincense, 3 plants of Sabine herb, 3 leaves of rue, 3 stalks of bryony, 3 white beans, 3 live coals, and 3 pints of wine. You must gather, macerate, and administer all these while standing, and he who administers the remedy must be fasting. Administer to each ox for three days, and divide it in such a way that when you have administered three doses to each you will have used it all. See that the ox and the one who administers are both standing, and use a wooden vessel. If an ox begins to sicken, administer at once one hen's egg raw, and make him swallow it whole. The next day macerate a head of leek with a hemina of wine, and make him drink it all. Macerate while standing, and administer in a wooden vessel. Both the ox and the one who administers must stand, and both be fasting … Give the cattle medicine every year when the grapes begin to change colour, to keep them well. When you see a snake skin, pick it up and put it away, so that you will not have to hunt for one when you need it. Macerate this skin, spelt, salt, and thyme with wine, and give it to all the cattle to drink. See that the cattle always have good, clear water to drink in summer-time; it

is important for their health. (Cat.70–3 translated in Hooper and Ash 1934)

First, let us think about the ingredients that are used to create these remedies, compiled in Table 1.1. Unsurprisingly, most of the ingredients (10/16, dark shading) are what we could consider normal kitchen ingredients, available to every herdsman. Furthermore, some plants common to the Mediterranean (3/16, lighter shading) and a few somewhat special ingredients (3/16, lightest shading) are listed. To procure these special ingredients, however, would have been comparatively easy.

It seems clear that actual availability must have been an important factor during the establishment of these early veterinary traditions: if your diligent herdsman is meant to take care of most ailments themself, it doesn't make sense to compile recipes with expensive and unattainable ingredients. It can be assumed that many a herdsman would likely have some of these ingredients with them when out grazing the cattle, as Varro and Palladius both hint that a good herdsman should be expected to carry a kind of veterinary medical bag when herding the cattle (see Varr. 2.2.20 in Goetz 1929; Pall. Vet. Med. 3 in Rodgers 1975). The general availability of these ingredients indicate that Cato's recipes probably reflect everyday life practice: it seems unlikely that this text is a purely intellectual endeavour, only advising readers of Rome's ruling class in a theoretical way. Rather, it seems plausible to interpret this passage as shedding light on genuine herdsman–cattle relationships of the Roman period. Indeed, it is generally assumed that, although Cato's work is stylistically refined and addresses the upper class, it was born out of genuine hands-on engagement with agriculture (Rex 1998, 13–15; Reay 2005).

The three more special ingredients might well have been established through the somewhat murky oral traditions of Roman veterinary practice. The beginnings of Mediterranean veterinary tradition are thought to consist of magical incantations and ingredients with presumed inherent magical powers (Adams 1995, 20; Earl 1939; Jones 1957; Mezzabotta 2001, 142). However, these origins might reveal specifically interesting aspects of Roman human–cattle relationships. In this chapter, it is hypothesised that these three special ingredients are a tangible link to the origins of veterinary treatment. The link to these origins, however, becomes tangible in other regards as well, as will be explained below. To illustrate the specific ritual character of those three ingredients, their significance to Roman ritual practice is briefly outlined here.

Table 1.1 Ingredients listed by Cato (*De agri cultura*, chapters 70–3) as part of recipes for healing cattle.

Ingredient	Classification	Ingredient	Classification
salt	kitchen ingredient	spikes of leek	kitchen ingredient
leek leaves	kitchen ingredient	thyme	kitchen ingredient
laurel leaves	kitchen ingredient	Sabine herb	common plant
garlic	kitchen ingredient	rue	common plant
white beans	kitchen ingredient	bryony	common plant
wine	kitchen ingredient	frankincense	special ingredient
hen's egg	kitchen ingredient	burning charcoal	special ingredient
emmer wheat	kitchen ingredient	snake skin	special ingredient

In the Mediterranean, frankincense has a long-standing tradition of being used for burial rites and communicating with the gods, starting in early Mesopotamia, transitioning to the ancient Greeks, the Romans, and, indeed, still prevailing in this function today (Hünemörder 2006). It was also used to cleanse animals before sacrificing them and is therefore potently linked to purification rites (Siebert 1999, 14). Burning charcoal, on the other hand, is even more closely linked to, and ubiquitous in, virtually all Roman sacrifices. The *tripus*, a specific charcoal brazier, is probably one of the best-known ritualistic objects of the Roman world and featured in a variety of rituals and sacrifices (Siebert 1999, 14, 255). As Roman sacrifices are inherently linked with burning something on charcoal, it is logical to assume that the use of burning charcoal in creating a remedy would evoke strong associations of entering a communication with the gods. Snakes (in this case, a snake's skin), on the other hand, are strongly associated with cults and rituals that are linked to healing, saving, or the benign communication with the dead (Toynbee 1973, 223–36). The most prominent example of this might be found in the famous Aesculapian staff: through it, snakes still grace pharmacies and hospitals around the world, demonstrating even today the ancient connotation of healing that these animals carry. It should become apparent that the ingredients that were provisionally classified as 'special' are indeed strongly linked to purification (frankincense), reaching out to the gods (burning charcoal), and healing (snake's skin). They seem to complement the more mundane kitchen ingredients, linking them to the realm of the gods and charging them with additional healing and purification qualities.

The three common plants – Sabine herb, rue and bryony – are also noteworthy. Mezzabotta noted that they might have certain medical virtues, as, for example, bryony is still used by some people as herbal medicine (2001, 145). Be this as it may, this chapter is not the place to discuss whether three leaves of rue really could have a significant medical effect on a fairly large mammal. Rather interesting and, I would argue, more relevant, is the number three: an important number in Roman religion and numeric mystery. Consider just a few examples: the Capitoline Triad, the most important gods in the city of Rome (Laroche 1998); Latin rhetoric, commonly structured in three parts (Quint. 9.4.22 in Butler 1922); or the various political collegiums consisting of three, like the *tresviri monetales* (Hamilton 1969). The consistent use of three pieces of each of the twelve ingredients listed by Cato (70) should therefore be interpreted as another indicator of the strong ritualistic connotation that characterises Roman veterinary practice.

However, the largest proportion of the ingredients named above consists of things that Romans probably wouldn't consider to be magically powerful or medically potent. After all, they are very familiar ingredients, being part of a normal Roman diet (they are commonly used, for instance, in Apicius' cooking handbook; see Danneil 1911). These ingredients are not normally associated with fodder, but rather with human food, yet they form the biggest part of a recipe for keeping animals (cattle) healthy. The cared-for animal seems to become a sort of patient, maybe partially transgressing – at least for the caring herdsman – to the human world. In moments of care, it seems, human food can become fodder, thus linking the otherwise separate worlds of humans and animals. This separation in the Roman world view was often remarked upon, for instance in Hine (2006); however, see also Bodson (1983) for the argument that this separation was always characterised by interaction and transgression; or Bertocchi and Orlandini (2007) for the argument that the Romans largely followed Aristotle in seeing the human and animal worlds as one. In this regard, it would be interesting to see if similar ingredients are used in Cato's recipe book for treating himself and his family members, as Plutarch mentions that Cato indeed wrote such a book as well (Plut. *Cat. ma.* 23.4 in Perrin 1914). It is, however, lost to us.

This observation of transgression opens up new possibilities for interpreting human–cattle relationships during moments of medical care, understanding them as moments with equalising tendencies. Apart from the fact that most of the employed ingredients are normally associated with human food, this interpretation is somewhat affirmed if we also consider the way in which the herdsman would administer the remedy. In chapter 70, quoted earlier, Cato states the person administering the remedy must stand upright, as must the animal. Furthermore, the person administering the remedy needs to be fasting. Even though we find nothing likewise about the cattle in this passage, it can be assumed that the procedure was meant to mirror what is described directly afterwards (chapter 71), where both the herdsman and the cattle need to be standing upright and be fasting together. Surely the medical effect of the remedy will not be enhanced in relation to the emptiness of the herdsman's stomach, at least not in any scientifically measurable way. Rather, this ritualistic behaviour links again to the older traditions of Roman veterinary practice, forming – together with the above-explained nature of the ingredients – a sort of ritualistic performance. But what kind of ritual is glancing at us through the veil of Cato's advice? What are its original intentions?

It seems that the human-associated food, the shared fasting and standing of herdsman and animal are actually going hand-in-hand, hinting at the same intention that once possibly was at the heart of caring for animals. In moments of care, it seems, humans and animals meet each other on a more level playing field. Roman views of dominating the natural world collapse, as the cattle becomes truly domestic and is treated more like a household member, not a tool. But before we speculate too much about possible interpretations and analytical pathways arising from this observation, let us see whether this aspect of Roman veterinary practice is only a faint shadow of its ritualistic past, a remainder to be glimpsed only in Cato's archaic work, or if it proves itself to be more consistent and enduring. Let us have a look at what happened in the three centuries separating Cato and Columella, exemplified by quoting Columella:

> Lassitude [fatigue] and nausea also can often be dispelled if you force a whole raw hen's egg down the animal's throat when it has eaten nothing; then on the following day you should crush spikes of 'Cyprian' or ordinary garlic in wine and pour it into its nostrils. Nor are these the only remedies which make for health. Many people mix also a generous quantity of salt with the fodder; some grate white horehound in oil and wine; some infuse fibres of leek, others grains of frankincense, others savin and crushed rue in unmixed wine and give them these medicaments to drink. Many people use the stalks of white-vine (bryony) and the shells of bitter-vetch as a medicine for oxen; some crush a snake's skin and mix it with wine. Thyme crushed in sweet wine and squill cut up and soaked in water are also used as remedies. All the above-mentioned potions in doses of three heminae given daily for three days purge the bowel and renew the animal's strength by driving away its maladies. (Col. 6.4.2–3 translated by Ash 1941)

The conservative character of Roman veterinary recipes becomes immediately apparent. Undeniably, Cato's remedies were, to a certain degree, included in the more extensive work of Columella, illustrated through the similarity of the chosen passages of Cato (70–3): 12 of the 16 ingredients named by Cato appear three centuries later in Columella's work, with examples covering all three previously introduced categories (kitchen ingredient, common plant, special ingredient). Columella (6.4–19) lists many more ingredients of this kind. But although the need for letting the cattle fast is mentioned in three cases (Col. 6.6.1; 9.1; 10.2 in Ash 1941), we don't find the same requirement for the herdsmen administering the treatment, which is equally true for the need to stand upright. This could be an indicator of the later changes in veterinary practice during the early imperial period, gradually moving away from its predominantly ritualistic past, maybe towards a more practical (but not necessarily better) approach. Even the number three within the recipes, although still in use (Col. 6.5.3; 6.3–5; 17.7 in Ash 1941), appears now alongside multiple other numbers and does not seem to hold the same importance any more as it did in Cato's advice.

The chosen passage of Columella clearly demonstrates the incorporation and ongoing relevance of older veterinary recipes, such as those found in Cato's work. However, it also becomes evident that certain aspects of Roman veterinary practice had lost importance in the interval. With ongoing time, it seems, the field moved gradually further away from primarily ritual activity. However, this makes it all the more interesting to look at what was lost. What can we learn about the beginnings of animal care in the Roman world? Which kind of human–animal relationships had once been present in Roman everyday life? Were they lost forever? Or did they reemerge in different contexts? Could it be that they never really disappeared, but remained one of the lenses through which animal feeding can and should be interpreted?

Unfolding human–cattle relationships

Roman veterinary practice, especially its everyday application, can best be understood as a convolute of different aspects. While, with the ongoing Roman period, the discipline seems to have changed and therefore lost certain aspects, the old ritualistic characteristics of it are still palpable. These traces have the potential to illuminate non-utilitarian, previously neglected aspects of animal feeding, not only applicable to the Roman period but to other times as well. Therefore, it seems irrational to sum up all these 'other' aspects as being murky, ritualistic (which seems to be equated with being irrelevant) traces of an uninteresting past (as for instance in Adams 1995, 20). This is not the only case in which ritualistic (or emotional) aspects of historic practices are dismissed as uninteresting and irrelevant, applying a truly post-Enlightenment point of view (see Jones et al. 2016 for discussion of how dismissing these aspects hinders meaningful historic interpretations). Let us instead speculate about some specific aspects of human–animal relationships that unfold themselves again and again in animal feeding, becoming palpable in precisely these 'unprofessional' aspects of feeding.

Surprisingly, the choice to feed animals with things that are not part of their natural diet – consider, in our case, the kitchen ingredients especially – is made by humans in all kinds of periods and regions, with different motivations and results. Some examples can be found in Chapters 2, 4, 6 and 10 of this book. Especially closely related is the case study from Chapter 2, describing the administration of this kind of food to raptors for veterinary purposes. Keeping this in mind, and considering the strongly human-associated qualities of the chosen ingredients in both Cato's and Columella's works, has implications for modern human–animal studies as well as anthropological, historical, and archaeological interpretations.

Consider, for example, studies of stable isotopes in animal remains. Nitrogen stable isotopes isolated from archaeological pet remains – especially cats and dogs – are regularly interpreted (in relation to the diet of humans from the same site) as hinting towards closeness and domestic relationships (Krajcarz et al. 2020; Guiry 2012; Edwards et al. 2017; Losey et al. 2020). But to the author's knowledge this has never been attempted with cattle or other farm animals. For these animals, divergent values in nitrogen and other stable isotopes are commonly interpreted not as possible hints of emotional closeness and intricate relationships of care, but exclusively in terms of either changing fodder regimes (Nitsch et al. 2017; Balasse and Tresset 2002; Groot et al. 2021) or the seasonal movement of livestock (Iorga et al. 2021; Müldner and Frémondeau 2021). Consideration of this aspect of feeding farm animals does not mean that it will prove feasible to discover traces of emotional closeness between cattle and their herdsman through analysing changing nitrogen stable isotope values, as there are many other factors impacting dietary values and feeding choices. However, in light of the practices of Roman veterinary feeding explored in this chapter, and considering additional emotional incentives for feeding human-associated food to animals – like those discussed in Chapters 2, 6 and 11 – it seems rash to interpret, for example, stable isotopes in livestock in so radically different a way from how they are interpreted in pet animals. The exclusive focus on utilitarian motivations as drivers for changing isotopic ratios, I argue, likely misses multiple other interpretational pathways linked to non-utilitarian drivers. Therefore, we should think together about how we can use the observations made in this chapter to ask new questions in archaeological sciences and other disciplines that deal with human–animal relations in the past.

Although feeding for veterinary purposes can be considered utilitarian up to a certain degree, a closer look reveals other motivations. In our case, the herdsman, arguably emotionally bonded to the cattle, might not cure them by feeding them a handful of herbs. But the fact that they take three of each herb while hungering together with their cattle and administering the potion in a certain shared position, may create a ritual that is fit to evoke the feeling of having tried: the empathic bond between herdsman and animal becomes more tangible, for both of them. The emotional satisfaction of trying to help, of empathising with the patient, might in fact be a major incentive for trying to heal the cattle through feeding in the first place, turning the seemingly utilitarian motivation onto its head and into a non-utilitarian, predominantly affective one. This observation of using utilitarian motivations as a rational cover for satisfying affective feelings linked to animal feeding is in no way confined to Roman period veterinary treatment but can be observed again and again, with further examples discussed in Chapters 6, 7 and 11. But the emotional relation enfolded in these moments of care is likely to have a significant impact on the animal's health in itself: there are numerous papers discussing the relevance of considering emotions in healthcare for humans, arguing that important factors of efficient healthcare are lost if viewing these interactions as purely rational and physiological processes (Heyhoe et al. 2016; Jiménez-Herrera et al. 2020); furthermore, learning and displaying empathy is considered a key quality in any modern medical education (Anfossi and Numico 2004; Hirsch 2007; Elam 2000). There is no reason to assume that this does not apply to some extent to veterinary medicine as well. The shared ritual and empathic care that the animal experiences might aid the healing process regardless of the precise chemical structure of the medicine itself. We might thus consider whether these 'unprofessional' aspects of historical animal care might inform and enrich our modern approach to veterinary practices.

It is also worth noting that the herdsmen administering the treatment not only satisfy their emotional needs through these ritualistic forms of feeding: there is a direct impact of the animal upon the herdsman. As they enter a more-or-less horizontal relationship during this shared ritual, the herdsman will surely become even more attached to the animals. The impact of treating an animal as bonded to oneself, especially noticeable in the invisible connection created by the equalising, ritualistic aspects of the administration, can be far-reaching, and we might ask ourselves about the importance of these bonds for the people of that time. This impact of animals on humans during ritualistic forms of feeding can be observed in modernity as well, as for example discussed in Chapters 6 and 7 of this book.

Conclusion

This chapter demonstrates that we can make more of the available material, be it written account, archaeological evidence, or anthropological study: once we start looking for new pathways of interpretation about why and how people were and are feeding non-pet animals beyond utilitarian purposes, new questions for academic debate emerge. Who can tell just how important the feeling to have tried really was for the herdsman himself? And we should also consider this in relation to the owner of the cattle, who is often not the same person: how important was the conducting of such veterinary rituals in order for the herdsman not to be blamed for any subsequent loss of livestock, and thus contribute to the ongoing relationship between owner and herdsman? Can we follow this pathway to ask even more questions about the relationship between herdsman and owner, maybe through considering the changes over time as well? Would the later application of remedies with less ritual (but still consisting of mostly impotent ingredients) fulfil the same role? And what about the alleged distinction between the human and animal realm in the Roman world view? Can these rituals, still tangible in moments of intricate care, hint at more nuanced developments, maybe pointing towards a gap between the world view of the ruling class and the everyday world perception of a common herdsman? Might the gradual loss of ritualistic modes of administering remedies itself be a hint of changing social relationships?

These are just some of the questions we have previously failed to ask by exclusively focusing on utilitarian livestock feeding. It is time to acknowledge the rich potential of also considering non-utilitarian motivations. Consider, for instance, the potential implications for the fields of stable isotope analysis, modern veterinary practice or numerous other disciplines analysing human–animal relationships. This chapter does not imply that any final answers or even relatively firm interpretations have yet been found. It merely demonstrates some of the new and exciting possibilities that we can gain through asking these questions and reconsidering the available material. Further possibilities, revealing other previously un-asked questions, will emerge as we follow the succeeding chapters into other parts of the Eurasian world and history. Let us use these as incentives to open new debates and reconsider old ones.

References

Adams, James Noel. 1995. *Pelagonius and Latin Veterinary Terminology in the Roman Empire*. Leiden: Brill.
Albarella, Umberto, Cluny Johnstone and Kim Vickers. 2008. 'The development of animal husbandry from the Late Iron Age to the end of the Roman period: A case study from south-east Britain', *Journal of Archaeological Science* 35 (7): 1828–48.
Anfossi, Maura and Gianmauro Numico. 2004. 'Empathy in the doctor–patient relationship', *Journal of Clinical Oncology* 22 (11): 2258–9.
Ash, Harrison Boyd. 1941. *Lucius Junius Moderatus Columella on Agriculture*. Cambridge, MA: Harvard University Press.
Balasse, Marie and Anne Tresset. 2002. 'Early weaning of neolithic domestic cattle (Bercy, France) revealed by intra-tooth variation in nitrogen isotope ratios', *Journal of Archaeological Science* 29 (8): 853–9.
Baldwin, Barry. 1963. 'Columella's sources and how he used them', *Latomus* 22 (4): 785–91.
Bertocchi, Alessandra and Anna Orlandini. 2007. In *La Médecine Vétérinaire Antique*, edited by Marie-Thérèse Cam, 169–78. Rennes: Presses universitaires de Rennes.
Bodson, Liliane. 1983. 'Attitudes toward animals in Greco-Roman antiquity', *International Journal for the Study of Animal Problems* 4 (4): 312–20.
Butler, Harold Edgeworth. 1922. *Quintilian. With An English Translation*. Cambridge, MA: Harvard University Press.
Danneil, Eduard. 1911. *Apicius Caelius: Altrömische Kochkunst in zehn Büchern*. Leipzig: Kurt Däweritz.
Diederich, Silke. 2007. *Römische Agrarhandbücher zwischen Fachwissenschaft, Literatur und Ideologie*. Berlin: De Gruyter.
Earl, Guthrie. 1939. 'History of veterinary medicine', *Iowa State University Veterinarian* 2 (1): 1–6.
Edwards, Richard, Robert Jeske and Joan Brenner Coltrane. 2017. 'Preliminary evidence for the efficacy of the canine surrogacy approach in the Great Lakes', *Journal of Archaeological Science: Reports* 13: 516–25.
Elam, Carol. 2000. 'Use of "emotional intelligence" as one measure of medical school applicants' noncognitive characteristics', *Academic Medicine* 75 (5): 445–6.
Fischer, Klaus-Dietrich. 1979. 'Palladius, *De vet. Med.* 14. 15, 5', *Liverpool Classical Monthly* 4 (4): 73–4.
Fischer, Klaus-Dietrich. 2020. 'Veterinary medicine'. In *The Encyclopedia of Ancient History*, edited by Roger S. Bagnall, Kai Brodersen, Craige Brian Champion, Andrew Erskine and Sabine R. Huebner, 1–5. Hoboken: Wiley-Blackwell.
Fischer, Klaus-Dietrich. 2023. 'Da ist der Wurm drin! Streuüberlieferung lateinischer veterinärmedizinischer Rezepte in frühmittelalterlichen humanmedizinischen Handschriften'. In *Documenta Archaeobiologiae, Animals and Humans through Time and Space: Investigating diverse relationships. Essays in Honour of Joris Peters*, edited by Nadja Pöllath, Nora Battermann, Stephanie Emra, Veronika Goebel, Ptolemaios Dimitrios Paxinos, Martina Schwarzenberger, Simon Trixl and Michaela Zimmermann, 125–38. Rahden: Marie Leidorf.

Fischer, Klaus-Dietrich, Diethard Nickel and Paul Potter. 1998. *Text and Tradition: Studies in ancient medicine and its transmission*. Leiden: Brill.
Fudge, Erica. 2013. 'Milking other men's beasts', *History and Theory* 52 (4): 13–28.
Goetz, Georg. 1929. *M. Terentius Varro, De Re Rustica*. Leipzig: Teubner.
Grau-Sologesta, Idoia, Maaike Groot and Sabine Deschler-Erb. 2022. 'Innovation and intensification: The use of cattle in the Roman Rhine region', *Environmental Archaeology*. https://doi.org/10.1080/14614103.2022.2090094.
Groot, Maike, Umberto Albarella, Jana Eger and Jane Evans. 2021. 'Cattle management in an Iron Age/Roman settlement in the Netherlands: Archaeozoological and stable isotope analysis', *PLOS ONE* 16 (10): e0258234.
Guiry, Eric J. 2012. 'Dogs as analogs in stable isotope-based human paleodietary reconstructions: A review and considerations for future use', *Journal of Archaeological Method and Theory* 19: 351–76.
Hamilton, Charles D. 1969. 'The tresviri monetales and the republican cursus honorum', *Transactions and Proceedings of the American Philological Association* 100: 181–99.
Heurgon, Jacques. 1976. 'L'agronome carthaginois Magon et ses traducteurs en latin et en grec', *Comptes rendus des séances de l'Académie des Inscriptions et Belles-Lettres* 120 (3): 441–56.
Heyhoe, Jane, Yvonne Birks, Reema Harison, Jane K. O'Hara, Alison Cracknell and Rebecca Lawton. 2016. 'The role of emotion in patient safety: Are we brave enough to scratch beneath the surface?', *Journal of the Royal Society of Medicine* 109 (2): 52–8.
Hine, Harry M. 2006. 'Rome, the cosmos, and the emperor in Seneca's Natural Questions', *Journal of Roman Studies* 96: 42–72.
Hirsch, Elliot. 2007. 'The role of empathy in medicine: A medical student's perspective', *Virtual Mentor* 9 (6): 423–7.
Hooper, William Davis and Harrison Boyd Ash. 1934. *Marcus Porcius Cato on Agriculture, Marcus Terentius Varro on Agriculture*. Cambridge, MA: Harvard University Press.
Hünemörder, Christian. 2006. 'Weihrauch'. In *Der Neue Pauly*, edited by Hubert Cancik, Helmuth Schneider and Manfred Landfester. Accessed 16 November 2023. http://dx.doi.org/10.1163/1574-9347_dnp_e12209480.
Iorga, Anastasia, Chris Gosden, Gary Lock and Rick Schulting. 2021. 'Stable carbon and nitrogen isotope analysis and Romano-British animal management along the Ridgeway, Oxfordshire', *Journal of Archaeological Science: Reports* 40: 103254.
Jacoby, Karl. 1885. *Dionysius Halicarnassus, Antiquitates Romanorum Quae Supersunt*. Leipzig: Teubner.
Jiménez-Herrera, María, Mireia Llauradó-Serra, Sagrario Acebedo-Urdiales, Leticia Bazo-Hernández, Isabel Font-Jiménez and Christer Axelsson. 2020. 'Emotions and feelings in critical and emergency caring situations: A qualitative study', *BMC Nursing* 19: 60.
Jones, Richard, Holly Miller and Naomi Sykes. 2016. 'Is it time for an elemental and humoral (re)turn in archaeology?', *Archaeological Dialogues* 23 (2): 175–92.
Jones, William Henry Samuel. 1957. 'Ancient Roman folk medicine', *Journal of the History of Medicine and Allied Sciences* 12 (10): 459–72.
King, Anthony. 1978. 'A comparative survey of bone assemblages from Roman sites in Britain', *Bulletin of the Institute of Archaeology University of London* 15: 207–32.
King, Anthony. 2001. 'The Romanization of diet in the western empire: comparative archaeozoological approaches'. In *Italy and the West: Comparative issues in Romanization*, edited by Simon Keay and Nicola Terrenato, 210–23. Oxford: Oxbow.
Krajcarz, Magdalena, Maciej T. Krajcarz, Mateusz Baca, Chris Baumann, Wim Van Neer, Danijela Popović, Magdalena Sudoł-Procyk, Bartosz Wach, Jarosław Wilczyński, Michał Wojenka and Hervé Bocherens. 2020. 'Ancestors of domestic cats in Neolithic Central Europe: Isotopic evidence of a synanthropic diet', *Proceedings of the National Academy of Sciences* 117 (30): 17710–19.
Laroche, Roland A. 1998. 'Proper names in the *Aeneid*: Their mystical numerical dimension', *Pallas* 48: 145–56.
Losey, Robert J., Eric J. Guiry, Tatiana Nomokonova, Andrei Gusev and Paul Szpak. 2020. 'Storing fish? A dog's isotopic biography provides insight into Iron Age food preservation strategies in the Russian Arctic', *Archaeological and Anthropological Sciences* 12 (200): 1–12. https://doi.org/10.1007/s12520-020-01166-3.

Mahaffy, John Pentland. 1889. 'The work of Mago on agriculture', *Hermathena* 7 (15): 29–35.
Mezzabotta, Margaret R. 2001. 'Ethnoveterinary treatments in Roman antiquity: Cato the elder's veterinary remedies', *Acta Classica: Proceedings of the Classical Association of South Africa* 44 (1): 137–52. https://journals.co.za/doi/pdf/10.10520/EJC27149.
Miller, Walter. 1913. *M. Tullius Cicero, De Officiis*. Cambridge, MA: Harvard University Press.
Müldner, Gundula and Delphine Frémondeau. 2021. 'Exploring Exeter's hinterland-relationships through its meat supply: the isotopic evidence in the Roman period'. In *Roman and Medieval Exeter and Their Hinterlands, from Isca to Exeter*, edited by Stephen Rippon and Nils Holbrook, 76–82. Oxford: Oxbow.
Nitsch, Erika, Stelios Andreou, Aurélien Creuzieux, Armelle Gardeisen, Paul Halstead, Valasia Isaakidou, Angeliki Karathanou, Dimitra Kotsachristou, Daphne Nikolaidou, Aikaterini Papanthimou, Chryssa Petridou, Sevi Triantaphyllou, Soultana M. Valamoti, Anastasia Vasileiadou and Amy Bogaard. 2017. 'A bottom-up view of food surplus: Using stable carbon and nitrogen isotope analysis to investigate agricultural strategies and diet at Bronze Age Archontiko and Thessaloniki Toumba, northern Greece', *World Archaeology* 49 (1): 105–37.
Ortoleva, Vincenzo. 1999. *Publii Vegetii Renati, Digesta Artis Mulomedicinalis, Liber Primus*. Catania: Sileno.
Ortoleva, Vincenzo. 2008. 'Vegetius Renatus'. In *The Encyclopedia of Ancient Natural Scientists: The Greek tradition and its many heirs*, edited by Paul Keyser and Georgia Irby-Massie, 823–4. London and New York: Routledge.
Perrin, Bernadotte. 1914. *Plutarch's Lives. With an English Translation*. Cambridge, MA: Harvard University Press.
Peters, Joris. 1998. *Römische Tierhaltung und Tierzucht: eine Synthese aus archäozoologischer Untersuchung und schriftlich-bildlicher Überlieferung*. Rahden: Verlag Marie Leidorf.
Reay, Brendon. 2005. 'Agriculture, writing, and Cato's aristocratic self-fashioning', *Classical Antiquity* 24 (2): 331–61.
Rex, Hannelore. 1998. 'Die Lateinische Agrarliteratur von den Anfängen bis zur frühen Neuzeit', doctoral thesis, Wuppertal: Bergische Universität.
Rodgers, Robert. 1975. *Palladii Rutilii Tauri Aemiliani Viri Inlustris, Opus Agriculturae, De Veterinaria Medicina, De Insitione*. Leipzig: Teubner.
Siebert, Anne Viola. 1999. *Instrumenta Sacra*. Berlin: De Gruyter.
Sykes, Naomi. 2014. *Beastly Questions: Animal answers to archaeological issues*. London: Bloomsbury.
Toynbee, Jocelyn Mary Catherine. 1973. *Animals in Roman Life and Art*. Ithaca, NY: Cornell University Press.
Viré, Ghislaine. 2007. 'La langue de la Mulomedicina de Végèce: tradition et innovation'. In *La Médecine Vétérinaire Antique*, edited by Marie-Thérèse Cam, 211–19. Rennes: Presses universitaires de Rennes.
Walker, R. E. 1973. 'Roman veterinary medicine'. In *Animals in Roman Life and Art*, edited by Jocelyn Mary Catherine Toynbee, 301–43. Ithaca, NY: Cornell University Press.

2
Food for flight: feeding of captive raptors in medieval Britain

Hannah Britton and Arthur Redmonds

Introduction

While much of the current academic research on raptors focuses on the pageantry and prestige associated with falcons as a sporting medium, both the pastoral and utilitarian aspects of food consumption and capture within falconry often remain disregarded (Pluskowski 2018, 150). However, there is a need to recognise the entanglement of utilitarian and non-utilitarian feeding in human–raptor relationships, and how the mutual feeding of falcon and falconer was embedded within a network of social relations and mutual agency through four 'construction' stages: the procurement of the bird, the training, the maintenance and finally hunting with the falcon itself. Using a framework similar to the process seen within the product-focused theoretical framework of the *chaîne opératoire* (operational chain; Sellet 1993), this chapter systematically explores the life history of the falcon, as the special status assigned to falcons makes them both 'product' and being, a tool and participant within hunting, forged through a chain of training and care (Bednarek 2018). For instance, the very nature of falconry inherently concerns raptors' predatory attributes, with the definition of 'falconry' (or 'hawking', used interchangeably) specified as 'taking quarry in its natural state and habitat by means of trained birds of prey' (International Association for Falconry 2018). Moreover, raptors are classified by their hypercarnivorous physiology: their long talons for grasping quarry and their sharp curved beaks used to rip and tear flesh. Hence, although medieval falconry is often seen as a means to a meal, to 'fill the bag' (Horobin 2011 pers. comms. quoted in Jacobi 2013, 477), room remains to debate the sociocultural underpinnings of feeding itself and, in particular, the degree to which interpersonal relationships were formed between handler and bird, within which nourishment was the currency of exchange and a reinforcement of hierarchy. Therefore, this chapter provides an interdisciplinary examination of the eighth- to sixteenth-century medieval falconer in Britain, drawing upon historic texts, archaeological evidence, personal experience with falconry and anthropological theory.

In Britain, falconry was thought to have arrived by the end of the eighth century, continuing until today with a temporary decline in popularity during the early modern period (Dobney and Jaques 2002; Grassby 1997; Carrington 1997, 462–4). Falconry birds appear within the archaeological record in Britain, with one identifying factor of a falconry bird (as opposed to a wild bird on an archaeological site) being the presence of possible hunted species in similar contexts (see Prummel 1997; Poole 2018). However, the reconstruction of feeding practices from archaeological remains is difficult as the everyday furniture of the falconer is largely absent, and the actual consumption of these prey species is suspect (Cherryson 2002, 308; Poole 2018, 1041; Serjeantson 2023, 1041; Yalden and Albarella 2009, 135–9). While there is current research being undertaken to explore the diets of these recovered falconry birds using biomolecular analyses (Britton in prep.), the feeding relationship of the sport must be largely recreated from the historical record.

The historical and archaeological evidence both suggest that by the later medieval period, falconry was largely the preserve of noble and ecclesiastical men and women, both in Britain (Oggins 2004; Mileson 2018, 387) and across Europe (Oggins 1986; Müller 1993; Bochenski et al. 2016; Lie 2018). Although subaltern individuals made up much of the staff that enabled the hunt, the falconers themselves produced the few surviving accounts, and it is from these high-status and literate circles that our documentary record of falconry remains (see Oggins 2004 for the English kings). The most pertinent sources are surviving medieval treatises on falconry, recording knowledge that would have been transmitted via inherited practice and understanding (Van den Abeele 2018). Although few remain, and cannot encompass the substantial corpus of now lost works, the shared methods and ideas indicate a strong continuity of practices through the medieval period. This chapter will draw examples from four treatises, whose more extensive survival facilitates our understanding of feeding:

1. Adelard of Bath's twelfth-century *De avibus tractatus* (translated in Burnett 1998) which focuses on the medical aspects of hawk rearing,

regarded as the earliest surviving English work on falconry and drawing from 'King Harold's Books' (Leggatt 1949, 138).
2. Holy Roman Emperor Frederick II von Hohenstaufen's thirteenth-century work *De arte venandi cum avibus* (translated in Wood and Fyfe 1943), a work born of the emperor's personal experience within his numerous parks and hunting lodges, and circulated throughout Britain (Egerton 2003, 41–2; Oggins 2004, 40; Caiola 2009, 84).
3. Theologian and natural philosopher Albertus Magnus' thirteenth-century *De animalibus* (translated in Kitchell and Resnick 2018), specifically the twenty-third book, which itself draws extensively from the works of 'William the Falconer' (King Roger II of Sicily's personal falconer), Dancus Rex, Aristotle, and from oral tradition of Frederick II's falconers (Egerton 2003, 89; Resnick and Kitchell 2022, 176; Oggins 1980).
4. Dame Juliana Berner's fifteenth-century *Boke of St Albans* (translated in Hands 1975), an English guide to the elite trappings of heraldry, hunting and hawking. Its influence is due to its popularity coinciding with the emergence of printing, seeing it republished twenty-two times by the early-seventeenth century, perpetuating medieval ideas far beyond the period (Grassby 1997, 39).

While globally a broad range of species is used for falconry – particularly eagles in the Asian Steppe – the scope of this chapter will be those species used in medieval Britain. These are the 'true' hawks, the sparrowhawk (*Accipiter nisus*) and goshawk (*Accipiter gentilis*), and the 'true' falcons, namely the peregrine falcon (*Falco peregrinus*), merlin (*Falco columbarius*), kestrel (*Falco tinnunculus*), hobby (*Falco subbuteo*) and the rare winter-visiting gyrfalcon (*Falco rusticolus*) (Serjeantson 2023, 157–70). The non-native lanner (*Falco biarmicus*) and saker (*Falco cherrug*) falcons have also been recorded historically as key falconry species used during the medieval period (Oggins 2004; Britton in prep.). Scavenging raptors such as kites and buzzards were rarely used in the sport and will not be discussed here, however the non-utilitarian human–raptor feeding relationships of red kites are explored in Chapters 4 and 11. While falconry birds fall generally between the two families of hawk or falcon, encompassing many differing hunting strategies, the terms will be used interchangeably throughout this chapter, and both held almost interchangeable roles within social hierarchies.

On the procurement and early care of falcons[1]

From its earliest stages, the strong relationship between the falconer and their birds is constructed through feeding, structured to emulate parental and hierarchical ideals. The capture of the bird has to be timed to allow for falconers to assume mastery of the bird without becoming a surrogate parent. Within this carefully planned relationship, the form and timing of food provision is key. Such captures require an intricate understanding of the natural feeding regimes inherent in the raising of the chicks and the stage at which parental relationships were formed and strengthened.

Unlike the present day, where captive breeding is viable, historical raptor populations were maintained primarily through the trapping of live birds (Cade and Berry 2018, 195–6; Hooke 2015, 272). Frederick II describes five methods to secure falconry birds (Wood and Fyfe 1943, 128–9), with the most common mode of procurement being the capture of juvenile birds. This usually consisted of a falconer climbing a tree or being lowered down a cliff face, depending on the species, and placing the eyass (young chick) gently into a basket to carry home, usually when the parent was absent from the nest (Wood and Fyfe 1943, 129).

Most authorities (Adelard, in Burnett 1998, 241; Frederick II, in Wood and Fyfe 1943, 128–9) agreed that falconers should delay capturing the eyasses after hatching to ensure that the bird has become fully formed and that parents have provided essential early nourishment, which was perceived as healthier, and difficult to replicate in captivity. The delay in retrieving the eyass also served to help it assimilate to other birds and to avoid early imprinting on the human, which can occur rapidly with hand-feeding and rearing (Frederick II in Wood and Fyfe 1943, 129). While imprinting is used by modern falconers as a means of training (particularly birds used in artificial insemination for captive breeding due to easier copulating; Blanco et al. 2009), it often comes with the negative association of the falconer with food, leading to continuous calling-out from the bird to be fed (Parry-Jones 2003, 96–102). Therefore, by delaying the time the falcon is retrieved from the nest, the medieval falconer can then introduce themselves as the fount of feeding more productively – ensuring the loyalty and malleability of the young falcon without establishing complete dependency, and therefore preserving individuality. Frederick II reasons, 'the longer they are fed by their parents the better and stronger will be their limbs and pinions

---

[1] Section titles are in the style of Frederick II's treatise *De arte venandi cum avibus*.

[flight feathers] and they are less likely to become screechers or gapers' (Wood and Fyfe 1943, 129).

The historic sourcing of young eyasses was well-documented, with location of eyries greatly protected and even recorded as assets in the Domesday Book (Oggins 2004, 20–1, 51, 84). Therefore, the remains of juvenile birds in archaeological sites has been put forward as suggestive of falconry practice in that location, such as juvenile sparrowhawks recovered from Nantwich Castle and a possible juvenile goshawk from Windsor Castle (Britton in prep.; Fisher 1986; Baker 2010).

Although many birds were likely acquired from nests, other birds such as 'passagers' (migrating birds before moult) and 'haggards' (captured birds after migration/moult) were also sourced, particularly those that had been traded from other regions (Frederick II in Wood and Fyfe 1943, 195–202; Oggins 2004, 12; Mellor 2006, 7; Serjeantson 2009a, 318). Frederick II advocates transporting birds during the night, though if travelling during the day is required, the bird must be fed meat soaked in cold water and 'should be sprinkled repeatedly with cold water from the attendant's mouth' (Wood and Fyfe 1943, 190, 195). The gyrfalcon was one of these frequently imported species, and at present four gyrfalcon individuals have been identified from British archaeological sites, at the urban and coastal settlements of Southampton, Winchester, and Bermondsey, London (Serjeantson 2009a, 323; Coy 2009, 42–3; Serjeantson 2009b, 182; Bates 2001; Pipe et al. 2011). It is known from historical texts that many of these birds were sought from Iceland (Mehler et al. 2018; Vaughan 1982, 331), and gyrfalcons have been excavated from Icelandic trading settlements such as Gásir in Eyjafjörður, where remains from two birds were recovered (Harrison et al. 2008, 103).

In 1495 an Act declared that 'none shal bear any hawk of the breed of England', banning the collecting of endangered native birds for falconry, but allowing the sourcing of birds from elsewhere with a licence (see Grassby 1997, 59; Manning 1993, 65, 79). Therefore, by the later medieval period the importation of falconry birds would have been necessary to practise the sport, and during transportation the care and health of the birds would have been vital. Unfortunately, from the corpus of treatises analysed, there is little description of the feeding of falcons on long voyages at sea.

On the training of falcons

For training both juvenile and adult-captured birds, structured feeding is the most potent tool available to the falconer. The manipulation of food is used from an early age and shapes the relationship, acting as a mediation device between the bird and the falconer during the training process. It is expressly provisioned by falconers within the early training so that falcons lose 'their natural aversion' to humans, which was particularly important for newly acquired haggards or passagers (Frederick II in Wood and Fyfe 1943, 225). Often food is used heavily in the manning (taming) process, with Albertus suggesting: 'The first regimen is carried out by never feeding the falcon anywhere but on the hand' (in Kitchell and Resnick 2018, 1593).

Ultimately, this early training with food enticement is used to reliably recall birds to the fist, with many falcons initially trained in retrieval using a lure (Albertus in Kitchell and Resnick 2018, 1578; Cummins 2001, 201; Standley 2008, 200). Standard lures were often constructed from the wings of a prey species bound together by leather and string so that meat could be attached, creating the illusion of a live bird so that when presented (thrown in the air in English practice) the falcon will be recalled to the easy meal (Frederick II in Wood and Fyfe 1943, 227, 243–4). Live lures were also used, often a weak fowl or pigeon, though Frederick II highlights an Armenian practice whereby a piglet is disguised in the skin of a hare and let free to run. The falcon would practise capturing the slower mammal before being flown at the much quicker live hare, building confidence and skill (Frederick II in Wood and Fyfe 1943, 255). However different these lure practices are, common to each is the curation by the falconer of a form of trust, and reliance on the lure via food supplementation, allowing for the falcon to continue returning to the falconer during the hunt.

The food negotiations between falcon and falconer during the training process define the difference between captive and wild. Frederick II refers to the falcon returning to the lure as 'forget[ting] her wild ways' (in Wood and Fyfe 1943, 227; sexual dimorphism in falconry birds is discussed later in this chapter); Adelard (in Burnett 1998, 267) suggests feeding a bird a concoction of pork fat and wine, prepared in a linen bag, to reward a hawk recalled from flight, and which could also be used to tame a wild hawk. Therefore, if the bird hunts and returns to the lure, it is considered tamed. However, this can be a vacillating and exorable status as, in a description of a newly acquired falcon, Frederick II (in Wood and Fyfe 1943, 203) suggests 'If she is emaciated she must be better fed, but not too much, else she will be made wild again'. In short, a tamed falcon can become wild just as a wild falcon can be tamed, with both of these transitions mediated through the quality and quantity of food available to the falcon.

Falconers were acutely concerned with provisioning food correctly during training, both for the effectiveness of the training and for the bird's health. In fact, training and feeding practices had a bidirectional causative relationship during the rearing of falcons, as sometimes food could only be provided at certain times and in specific measures. One reason for this is crop stasis, or 'sour crop', which occurs when falcons become ill, often fatally, due to the fermentation of food in the crop (the front-facing pouch under the head in which the falcon stores food) as it has not been emptied into the stomach fast enough (Cooper 2008, 96). This sometimes occurs in young chicks where they have been overfed, or with adults when a substantial amount of food has been consumed in warm temperatures, and thus cannot be digested quickly enough before decaying. Medieval falconers were aware of this plight and often only provided food twice a day, when 'The heat of midday will have no chance to affect food remaining in the crop' (Frederick II in Wood and Fyfe 1943, 135–6).

Therefore, the time you feed a bird affects the time you will train with it, while the provision and denial of food is an integral part of the training itself. For instance, falcons were encouraged to gain weight in summer so that they could moult their feathers effectively, whereas leading up to the hunting season falcons were encouraged to lose fat in a regime of semi-starvation in order to establish an incentive to hunt (Albertus in Kitchell and Resnick 2018, 1595; Oggins 2004, 30). The bird must be at prime hunting weight to allow hunting effectively: 'One should therefore adjust a falcon's weight mainly to her eagerness to fly, taking care to maintain her strength' (Frederick II in Wood and Fyfe 1943, 249, 252–9). The importance of this seasonal dichotomy is why the sport is often considered to have been revolutionised with the widespread usage of falconry weighing scales in the early modern period, which allowed accurate calculations of flying and fat weights (Mellor 2006, 15).

On the maintenance and health of falcons

Similar to feeding in training, feeding fulfils equally essential roles in the health and care management of falconry birds in captivity, while perpetuating human dominance over the bird. Most sources discuss the best foodstuffs at length, and collectively agree that falconry birds should maintain a diet that mirrors a natural varied wild diet: 'The wise falconer ought to strive to follow nature in the art of feeding' (Albertus in Kitchell and Resnick 2018, 1579; Adelard in Burnett 1998, 265; Frederick II in Wood and Fyfe 1943, 426). One archaeological study demonstrated that almost all of the sixty-three sites analysed yielded falcon and hawk prey species (see Poole 2018), with ducks and other water birds particularly prevalent, although exact estimates are likely skewed according to preservation and excavation bias. There is also evidence of prey species purposely kept on site at high-status locations for hunting and the replenishment of deer parks: for instance, the thirteenth-century royal mews at Charing Cross had a 'house … to keep … cranes at which the gyrfalcons were flown' (Colvin 1963, 551).

Though easier to procure, domesticates – particularly domestic mammals over birds – were considered less healthy by the medieval falconers, although when wild foodstuffs could not be sourced, the healthiest domestic animals must be butchered for meat for the falcons (Frederick II in Wood and Fyfe 1943, 133; Adelard in Burnett 1998, 265). This sentiment opposes medieval notions that food consumed by the poor class mirrored food fed to animals (see Langdon 2018, 5–6; Dyer 1989, 57), with the foodstuffs eaten by the falcons suggested by these treatises likely superseding the quality of foods consumed by most social classes (Serjeantson 2009c; Sykes 2009, 163–6).

The medieval falconer was well aware of the impact of diet on bird health, and was conscious of dietary deficiency diseases such as rickets, and therefore food as sustenance was discussed intimately in the historical literature for this very reason (Adelard in Burnett 1998, 243; Olmos de León 2018, 545). Moreover, food sources were carefully inspected for diseases, particularly pigeons, which were often carriers of trichomoniasis (colloquially known as 'canker' or 'frounce' in falcons; Olmos de León 2018, 547). Foodstuffs were also provisioned as active health prescriptions and remedies for ailing birds. This parallels discussions in Chapter 1, which examines the use of 'human' foods as Roman cattle veterinary treatments. Adelard, arguably the leading authority on these English health practices, often describes these prescriptions in the format of instructions and recipes:

> Nephew: What do you think should be done for an obstructed gut?
> Adelard: Give him large pieces of cow's meat that have been warmed in water until they become white and give them to him. (Adelard in Burnett 1998, 261)

If they became ill, the falcons were occasionally treated to a rich diet full of red meats and other foods familiar upon the lord's table. Albertus (in Kitchell and Resnick 2018, 1619) for instance prescribes 'meat smeared with honey' for a bird thirsty from fever, the next round of treatment being smeared instead with 'cold oil treated with roses'. Even more exorbitant, Adelard also prescribes human by-products in remedies, such as human excrement (Burnett 1998, 251) or, when he describes the prescription for mites, human skeletal remains:

> Take the tooth of a hanged man, or that bone which is attached to the larger bone of the arm which they call the 'escinum', reduce it to powder and give it to him with his meat, and he will be saved. (Adelard in Burnett 1998, 257)

Cummins (2001, 209) additionally describes a Spanish practice that uses mummies in a remedy for ill falcons. This is particularly notable considering the cultural expression of food sharing, where some medieval attitudes towards consuming non-human foodstuffs were that it was dehumanising and sinful, believing that to share the diet of an animal was to take on its behaviours (Langdon 2018, 3–4). Therefore, the very thought of (mummified) humans as the 'meal' is a complete perversion of the natural order and the 'chain of being'.

From historical sources, it is clear that many falconers had a holistic understanding of food provisioning beyond mere sustenance, particularly considering food texture, temperature and quality beyond that necessary for baseline nutrition. Adelard suggests providing birds with 'roughage twice a week' (in Burnett 1998, 273), while Frederick II recommends heating the meat in warm water to simulate freshly killed prey (in Wood and Fyfe 1943, 134; see also Berners in Hands 1975, 24). Other contemporary sources, such as writing by the fourteenth-century Castilian Pero López de Ayala (in Cummins 2001, 203), recommend feeding the bird portions of meat with less butchery preparation so that the falcon must tug and rip at the flesh, providing a source of enrichment and exercise. On this point, Frederick II indirectly demonstrates that the human processing of prey carcasses for meals likely had a discernible impact on the captive birds (in Wood and Fyfe 1943, 134). The Vatican illuminated manuscript depicts the coping (trimming) of overgrown beaks and talons of the falconry birds, which can become overgrown in captive birds due to lack of abrasion usually achieved by consuming and hunting prey in the wild (Wood and Fyfe 1943, 132, plate 70), leading to splitting and becoming infected unless manually reshaped (Cooper 2008, 181). While not archaeologically identifiable, in modern populations this occurs most commonly (but not exclusively) after several years in captivity, and is likely due to the consumption of non-ideal textured foodstuffs, including particularly butchered portions rather than whole carcasses (Cooper 2008, 129; Lascelles 1971, 155–6; see also Chapter 9 for the effects of feeding on plastic morphological change in captive mammals).

Overall, correctly provisioning foodstuffs was the top priority for falconers and was likely the first response to assist ailing birds. The very nature of falconry – the hunt – was dangerous for falcons, with several archaeological specimens recovered from excavations demonstrating high levels of healed trauma and subsequent disease, as well as detailed historical records depicting many falcon injuries and fatalities (Britton in prep.; Oggins 2004, 55, 85, 104–5, 123). Life in captivity would also have been treacherous as, despite protection from the natural world, the close proximity of other birds significantly increases the transmission of disease and the spread of fungi. Therefore, the significant number of falconry treatises concerning health is unsurprising, a sentiment echoed by Cummins (2001, 208), who suggests: 'The craft and mystery of medieval falconry was pervaded by a vicarious hypochondria'. While the success of some of these remedies can be debated, the comprehensive knowledge of food provisioning cannot be denied, and arguably stands the test of time.

On the use of falcons in hunting

Up to this point the operational chain that dictates feeding and training has led towards the capture and killing of quarry. Once the hunt begins, however, the act of feeding becomes inverted and the falcon becomes the bearer of food, and the bearer of new social roles. The presence of ritual or social significance is evident in the almost superfluous nature of the hunt (Almond 2018). As previously argued, there is significant time involvement in the procurement, training and maintenance of these birds leading up to the actual hunt, and during this process the food provided has acted as an investment in the relationship, with very little return. This shows that falconry is less concerned with the acquisition of food for human subsistence, and is more generally focused on the entertainment and sport value (Judkins 2013).

The hunt is intertwined with ideas of ritual, where the killing of the animal is a significant sociocultural event that reinforces group identity (Sykes 2014, 156). This is particularly evident in the recommended hunting of the crane by Frederick II (in Wood and Fyfe 1943, 290), who suggests that once the prey is taken down, the falconer should cut open the breast of the crane and give the heart to the falcon, thus quickly killing the crane to stop any further danger to the falcon, and also ensuring that the falcon does not become soiled by blood. This process is reminiscent of both the curée practice conducted with dogs in hunting, and the ritual butchering 'unmaking' of the deer carcass (Cummins 2001; Langdon 2018, 8; Sykes 2014, 72–3; Sykes 2010; Salisbury 2011, 36–7).

In many cases the value of the hunt was also dependent on the prey being targeted: a larger, more extravagant prey demonstrated the prowess of the falcon and affirmed the ability of the owner to adequately feed and train their bird. Falcons were flown at species that are much larger than those they would usually attack in the wild, particularly the common crane (*Grus grus*), which is historically thought to have bred in Britain until the seventeenth century and was the most significant prey species for the gyrfalcon (Oggins 2004, 16; Serjeantson 2010, 148–9; Albarella and Thomas 2002; Yalden 2002; Poole 2018). The ascribed value of the raptor also reflected sex differences, with female birds valued over males in many cases due to their larger size (an instance of reverse sexual dimorphism), and larger raptors being able to take down significantly larger prey (Petrosillo 2023, 15–16; Prummel 1997, 336). Therefore the social separation of bird sexes was far more defined historically than today, with differing terminology adopted to define the sexes: male falcons were often referred to as 'tiercels', or 'muskets' in the case of sparrowhawks (Oggins 2004, 12; Serjeantson 2009a, 318).

However, in some early medieval falconry practices falcons were considered food providers, with the goshawk occasionally referred to as the 'cook' or 'kitchen' hawk (Almond 2018, 1121; Evans 1990, 84). Furthermore, falcons were sometimes framed as necessary hunting devices, rather than predominantly as exhibitions of noble prowess, as exemplified in Ælfric of Eynsham's *Colloquy*:

> Teacher: How do you feed your hawks?
> Fowler: They feed themselves and me in winter, and in the spring I allow them to fly off into the wood, and I catch the young hawks in autumn and tame them.
> Teacher: And why do you allow the tamed ones to fly away from you?
> Fowler: Because I don't want to feed them in summer, because they need to eat [too much]. (Translated in Lacey 2018, 1094; Oggins 2004, 44–5)

This indicates that during the early medieval period falcons were used as practical tools for subsistence, and discarded in summer when they became a drain on resources (Lacey 2018, 1094; Oggins 2004, 44–5). This sentiment of only valuing the profitable bird, either in food supplies or grandeur, was possibly more ubiquitous than treatises suggest, as many of the British archaeological bird remains were seemingly discarded without care. This strongly contrasts with the image of highly cherished and cared-for falcons that these treatises propose, as many individual birds were recovered from middens and cesspits, such as the partial skeletons from Windsor Castle and Greyhound Yard, Dorchester (Baker 2010; Maltby 1993, 333, 339). It could be argued that, in this aspect, the birds were not valued beyond their usage; this idea is reinforced in the early modern period, which saw the decline in falconry coinciding with the popularisation of shooting, and later the active persecution of wild raptor species as competitors for game (Grassby 1997, 52).

On trophic social hierarchies

Beyond the literal hunting, the exchange and provision of food was structured according not only to the needs of the bird, but to their perceived place in wider social hierarchies. Indeed, their species, age, sex and experience all afforded them a place in social structures often parallel to those experienced by their owners (Jones 2021, 445; Almond 2018, 1123–5). Certainly, birds are socially stratified in the *Boke of St Albans*, which defines what bird is appropriate for each social rank (Berners in Hands 1975, 54–5; Cummins 2001, 187–91; Yalden and Albarella 2009, 136). However, authorities on this topic agree that this hierarchical structure has been taken too seriously and was not stringently replicated in reality (Cummins 2001; Oggins 2004), for example Henry VIII was frequently associated with flying merlins, despite that species being deemed only appropriate for a Lady (Horobin 2004, 64–5).

It is undeniable that gyrfalcons – dubbed 'the king's bird' – were frequently sent to royalty in diplomatic conventions and were considered as the noblest of all birds (Frederick II in Wood and Fyfe 1943, 225; Oggins 2004, 13; Oppitz-Trotman 2010, 78). As the largest of the falcons found in Britain at this time, it was likely the gyrfalcons' rarity and ability to hunt cranes that granted it this unique social rank, which also allowed it larger food allowances (Frederick II in Wood and Fyfe 1943, 225; Oggins 2004, 12). This ranking, which puts the birds themselves in parallel to human societal order, is almost a reflection of trophic order, with the dominant consumers at the top of the chain (see Bartosiewicz 2012 for correlation to size). As such, the ordering of birds reflected ideas of social order within the medieval world, each drawing legitimacy from the other, as 'social differences were seen in accordance with the laws of nature' and as such were divinely ordained and sacrosanct (Duby 1997, 343–5).

The association between trophic levels and social order is compounded by the thematic use of falcons in literature, with birds being assigned values based on their dietary habits (see also Chapter 4 on the historic feeding of red kites in urban spaces, and Chapter 5 on the role of food in conflict between humans and corvids). As their diet often mirrors that of the nobility, they are considered 'noble'. This is directly opposing other ideas of human and animal food-sharing, namely that meat produced from a human kill is deemed human food, whereas prey killed by animals is animal food (Langdon 2018, 8; Salisbury 2011, 34–5). In this case, the falconry bird can be seen as an exception – as an extension of the hunter – and therefore, the work and the exchange of food prior to hunting allows for the food produced to be deemed righteous for human consumption (Sykes 2014, 111). Therefore, in addition to simply having the permission to engage in hunting in elite spaces and residences, the falcon is seen as a member of the hunt, and granted the social status that this accords. This is particularly evident for elite women, where hunting with a falcon allowed for the preservation of their femininity, as it was the falcon acting as a hunting partner actually making the kill (Sykes 2014, 110).

The importance of the trophic order is distinguished also by the treatment of carrion birds and other seemingly lesser species. Unlike those used in falconry, the scavenging species of kites and buzzards were not as fairly depicted in contemporary literature, and were sometimes the prey species of falconry birds (Wheater 2018, 144). The red kite specifically has been argued to have been the main prey for gyrfalcons and saker falcons in both England and France until the nineteenth century (Dobney and Jaques 2002, 17–18; Zeiler 2010, 166). On the other hand, eagles were considered greatly noble and valuable, but were only considered a 'novelty' to fly for medieval falconers (Frederick II in Wood and Fyfe 1943, 111; Berners in Hands 1975, 54–5; Oggins 2004, 115; Holmes 2018).

This relationship between feudal and trophic order is further emphasised by its repeated appearances in the imagined anthropomorphic depictions of literature, where the nobleman is often portrayed as a falcon or hawk (Evans 1990, 91–9). This trope is found frequently in literature, particularly in the twelfth-century fables of Marie de France, but also in Chaucer's *Parliament of Foules* (1382), which presents an equally stratified anthropomorphic depiction of raptors within which the nobility of the eagle is emphasised (Evans 1990, 86; Salisbury 1996, 57–8; Gutmann 2012). This notion of social rank in parallel to human order is neither new nor an unrecognised topic; on this point Cummins (2001, 189), in his renowned book *The Hound and the Hawk*, describes Bandello's Novellino as an example of this perennial metaphor. The king of Persia is depicted as finding that his prized falcon has killed an eagle, instead of the heron it was chasing; the barons and gentlemen 'held the falcon for one of the best in the world', yet the king withdrew the falcon, adorned it in a gold crown and ceremonially beheaded it the next day.

> The falcon, having, then, slain his queen and the queen of all the birds, who is there can with reason blame me if I have caused cut off his head? Verily, methinketh, no one. (*The Novels of Matteo Bandello*, translated in Payne 1890, 47–9)

As depicted in the *Boke of St Albans*, the eagle is seen as synonymous with an emperor or king, and therefore the killing is a violation of the natural order. Indeed, although a falcon is also a predator, it does not outrank an eagle – which is a true apex in its landscape.

The natural characteristics, wild connotations and social standing of falcons and hawks have also been used as metaphorical expressions of civility, where the anthropomorphised 'tame' hawk is considered refined and a 'wild' hawk is considered uncouth. In the case of the didactic writings of *Aviarium* by Hugh of Fouilloy (translated in Clark 1992), the contrasting literary imaginations of the taming of the hawk are used to embody a man learning monastic teachings of God (Wheater 2018). The allegorical representation of the birds again is defined by the food they consume, as detailed in historical treatises – the wild hawk preys on tame fowl and immediately eats it all, while the tame hawk hunts wild birds and preserves them for its master, being rewarded with the quarry's heart, while the innards soiled by excrement are thrown away (Clark 1992, 143–7). Therefore, in matters of social order and civility, we can see that sociocultural significance is assigned according to the perception of the natural world.

Conclusion

In all four 'construction' stages of a falconry bird – procurement, training, maintenance and use in hunting – feeding holds an integral role in establishing socio-hierarchical ideals across all historical treatises. From the first interaction to the manning process, and then training on the lure, food acts throughout as currency invested in the falcon, where the final product is both a functioning hunting device and a hunting partner. Foodstuffs are heavily scrutinised by medieval treatises, from quality to provisioning, with the quantity regulated closely in the diametrically opposed processes of fattening or starving the falcon, depending on seasonal needs. As a good falconry bird was considered one with good plumage and free from any ailments, a considerable number of treatises were concerned with best health practices. These remedies frequently came in the form of food provisioning, often described in a manner similar to recipes.

Falconry birds were frequently valued, both sentimentally and economically, according to the prey that they could capture, and as such there are perennial thematic undertones shared throughout historical treatises and literature regarding the human–falcon relationship, with diet at the focal point. These birds were frequently portrayed anthropomorphically, often with dominance over a landscape analogous to the domination of the food chain as a predator. Therefore, both in imagined landscapes as in reality, the falcon is equated with seigneurial power. On this point, the oscillation between 'wild' or 'tame' pivots upon the quality and quantity of food consumed, which – as argued by Frederick II – *never* makes a falcon domesticated. Ongoing archaeological research will yield further insights into the realities of falcon feeding using the theoretical underpinnings of feeding itself as essential to the understanding of falconry practice.

References

Albarella, Umberto and Richard Thomas. 2002. 'They dined on crane: Bird consumption, wild fowling and status in medieval England', *Acta Zoologica Cracoviensia* 45 (special issue): 22–38.

Almond, Richard. 2018. 'Hunting from the fist: looking at hawking and falconry in late medieval England (1000–1500) through art history'. In *Raptor and Human: Falconry and bird symbolism throughout the millennia on a global scale*, edited by Karl-Heinz Gersmann and Oliver Grimm, 1117–48. Kiel: Wachholtz Verlag.

Baker, Polydora. 2010. 'Procurement, presentation and consumption of domestic and wildfowl at Windsor Castle, England in the 12th–14th c.'. In *Birds in Archaeology: Proceedings of the 6th meeting of the ICAZ Bird Working Group in Groningen (23.8–27.8. 2008)*, edited by Wietske Prummel, Jørn Zeiler and Dick C. Brinkhuizen, 57–69. Groningen: Barkhuis.

Bartosiewicz, László. 2012. 'Show me your hawk, I'll tell you who you are'. In *A Bouquet of Archaeozoological Studies: Essays in honour of Wietske Prummel*, edited by D. C. M. Raemaekers, K. E. Esser, R. C. G. M. Lauwerier and J. T. Zeiler, 179–88. Groningen: Barkhuis.

Bates, Andrew. 2011. 'Animal and bird bone'. In *Trade and Prosperity, War and Poverty: An archaeological and historical investigation into Southampton's French quarter*, edited by Richard Brown and Alan Hardy, 223–33. Oxford: Oxford Archaeology.

Bednarek, Walter. 2018. 'Emotions and motivation of the falconer and his relationship with the trained raptor: attempt at an evolutionary-biological interpretation'. In *Raptor and Human: Falconry and bird symbolism throughout the millennia on a global scale*, edited by Karl-Heinz Gersmann and Oliver Grimm, 285–99. Kiel: Wachholtz Verlag.

Blanco, J. M., D. E. Wildt, U. Höfle, W. Voelker and A. M. Donoghue. 2009. 'Implementing artificial insemination as an effective tool for ex situ conservation of endangered avian species', *Theriogenology* 71 (1): 200–13.

Bochenski, Zbigniew M., Teresa Tomek, Krzysztof Wertz and Michał Wojenka. 2016. 'Indirect evidence of falconry in medieval Poland as inferred from published zooarchaeological studies', *International Journal of Osteoarchaeology* 26 (4): 661–9.

Britton, Hannah. In preparation. 'Flying through History: Human–raptor relationships in Britain', doctoral thesis to be submitted to University of Exeter.

Burnett, Charles (ed./trans.). 1998. *Adelard of Bath: Conversations with his nephew*. Cambridge: Cambridge University Press.

Cade, Tom J. and Robert B. Berry. 2018. 'The influence of propagating birds of prey on falconry and raptor conservation'. In *Raptor and Human: Falconry and bird symbolism throughout the millennia on a global scale*, edited by Karl-Heinz Gersmann and Oliver Grimm, 195–218. Kiel: Wachholtz Verlag.

Caiola, Amelia. 2009. 'An Exploration of Falconry and Hunting in the Middle Ages Based on the Work of Emperor Frederick II *De arte venandi cum avibus* and its Links to Science, Natural Philosophy and Literature', doctoral thesis, New York University.

Carrington, Ann. 1997. 'The horseman and the falcon: Mounted falconers in Pictish sculpture', *Proceedings of the Society of Antiquaries of Scotland* 126: 459–68.

Cherryson, Annia. K. 2002. 'The identification of archaeological evidence for hawking in medieval England', *Acta Zoologica Cracoviensia* 45 (special issue): 307–14.

Clark, Willene B. (trans.). 1992. *The Medieval Book of Birds: Hugh of Fouilloy's Aviarium*. Binghamton, NY: Medieval & Renaissance Texts & Studies.

Colvin, Howard M. 1963. 'Westminster Palace'. In *The History of the King's Works*, Volume 1, edited by Reginald Allen Brown, Howard M. Colvin and Alan John Taylor, 491–552. London: Her Majesty's Stationery Office.

Cooper, John E. 2008. *Birds of Prey: Health and disease*. Oxford: Wiley.

Coy, Jennie. 2009. 'Late Saxon and medieval animal bone from the western suburb'. In *Food, Craft, and Status in Medieval Winchester*, edited by Dale Serjeantson and H. Rees, 27–55. Winchester: Winchester Museums Service.

Cummins, John G. 2001. *The Hound and the Hawk: The art of medieval hunting*. London: Phoenix Press.

Dobney, Keith and Deborah Jaques. 2002. 'Avian signatures for identity and status in Anglo-Saxon England', *Acta Zoologica Cracoviensia* 45 (special issue): 7–21.

Duby, Georges. 1997. *The Three Orders: Feudal society imagined*. Translated by A. Goldhammer. Chicago: University of Chicago Press.

Dyer, Christopher. 1989. 'The aristocracy as consumers'. In *Standards of Living in the Later Middle Ages: Social change in England c.1200–1520*, by Christopher Dyer, 49–85. Cambridge: Cambridge University Press.

Egerton, Frank. 2003. 'A history of the ecological sciences, part 8. Frederick II of Hohenstaufen: Amateur avian ecologist and behaviorist', *Bulletin of the Ecological Society of America* 84 (1): 40–4.

Evans, Dafydd. 1990. 'The nobility of knight and falcon'. In *The Ideals and Practice of Medieval Knighthood*, Volume 3: *Papers from the fourth Strawberry Hill conference 1988*, edited by Christopher Harper-Bill and Ruth Harvey, 79–100. Woodbridge: Boydell Press.

Fisher, C. T. 1986. 'Bird bones from the excavation at Crown car park, Nantwich, Cheshire', *Circaea* 4 (1): 55–64.

Grassby, Richard. 1997. 'The decline of falconry in early modern England', *Past & Present* 157: 37–62.

Gutmann, Sara. 2012. 'Chaucer's chicks: Feminism and falconry in *The Knight's Tale, The Squire's Tale*, and *The Parliament of Fowls*'. In *Rethinking Chaucerian Beasts*, edited by Carolynn Dyke, 69–83. New York: Palgrave Macmillan.

Hands, Rachel (trans.). 1975. *English Hawking and Hunting in 'The Boke of St Albans': A facsimile edition of sigs a2–f8 of 'The Boke of St Albans' [1486]*. London: Oxford University Press.

Harrison, R., H. M. Roberts and W. P. Adderley. 2008. 'Gásir in Eyjafjörður: International exchange and local economy in medieval Iceland', *Journal of the North Atlantic* 1 (1): 99–119.

Holmes, Matilda. 2018. 'King of the birds! The changing role of white-tailed (*Haliaeetus albicilla*) and golden-eagles (*Aquila chrysaetos*) in Britain's past', *Archaeofauna* 27: 173–94.

Hooke, Della. 2015. 'Beasts, birds and other creatures in pre-conquest charters and place-names in England'. In *Representing Beasts in Early Medieval England and Scandinavia*, edited by Michael D. J. Bintley and Thomas J. T. Williams, 253–82. Woodbridge: Boydell & Brewer.

Horobin, David. 2004. *Falconry in Literature: The symbolism of falconry in English literature from Chaucer to Marvell*. Surrey, BC: Hancock House.

International Association for Falconry. 2018. 'International Association for Falconry and Conservation of Birds of Prey'. Accessed 27 January 2023. https://iaf.org/.

Jacobi, Leor. 2013. 'Jewish hawking in medieval France: Falconry, Rabbenu Tam, and the Tosafists', *Oqimta* 1: 421–504.

Jones, Peter J. A. 2021. 'Bones, fire, and falcons: Loving things in medieval Europe', *Journal of Material Culture* 26 (4): 433–50.

Judkins, Ryan. 2013. 'The game of the courtly hunt: Chasing and breaking deer in late medieval English literature', *Journal of English and Germanic Philology* 112 (1): 70–92.

Kitchell, Kenneth F. and Irven Michael Resnick (trans.). 2018. *Albertus Magnus On Animals V1 2: A medieval Summa Zoologica*. Revised edition. Columbus: Ohio State University Press.

Lacey, Eric. 2018. 'The charter evidence for falconry and falcon-catching in England and Wales, c. 600–c. 1100'. In *Raptor and Human: Falconry and bird symbolism throughout the millennia on a global scale*, edited by Karl-Heinz Gersmann and Oliver Grimm, 1089–116. Kiel: Wachholtz Verlag.

Langdon, Alison. 2018. 'Fit for a dog? Food sharing and the medieval human/animal divide', *Society & Animals* 28 (1): 1–17.

Lascelles, Gerald. 1971. *The Art of Falconry*. Newton Centre, MA: C. T. Branford Co.

Leggatt, N. J. Shirley. 1949. 'The *Book of St. Albans* and the origins of its treatise on hawking', *Studia Neophilologica* 22 (2): 135–45.

Lie, Ragnar. 2018. 'Falconry, falcon-catching and the role of birds of prey in trade and alliance gifts in Norway (800–1800 AD) with an emphasis on Norwegian and later foreign participants in falcon catching'. In *Raptor and Human: Falconry and bird symbolism throughout the millennia on a global scale*, edited by Karl-Heinz Gersmann and Oliver Grimm, 727–86. Kiel: Wachholtz Verlag.

Maltby, Mark. 1993. 'Animal bones'. In *Excavations at the Old Methodist Chapel and Greyhound Yard, Dorchester, 1981–1984*, edited by Peter J. Woodward, Susan M. Davies and Alan H. Graham, 315–40. Dorchester: Dorset Natural History and Archaeology Society.

Manning, Rodger. 1993. *Hunters and Poachers: A social and cultural history of unlawful hunting in England 1485–1640*. Oxford: Oxford University Press.

Mehler, Natascha, Hans Christian Küchelmann and Bart Holtermann. 2018. 'The export of gyrfalcons from Iceland during the 16th century: a boundless business in a proto-globalized world'. In *Raptor and Human: Falconry and bird symbolism throughout the millennia on a global scale*, edited by Karl-Heinz Gersmann and Oliver Grimm, 995–1020. Kiel: Wachholtz Verlag.

Mellor, Gordon T. 2006. 'Falconry in Britain between 1750 and 1927: The survival, organisation and, development of the sport', doctoral thesis, De Montfort University.

Mileson, Stephen. 2018. 'Royal and aristocratic landscapes of pleasure'. In *The Oxford Handbook of Later Medieval Archaeology in Britain*, edited by Christopher Gerrard and Alejandra Gutiérrez, 386–400. Oxford: Oxford University Press.

Müller, Hans-Hermann. 1993. 'Falconry in central Europe in the Middle Ages'. In *Exploitation of Wild Animals through Time*, Volume 13: *International meetings of archaeology and history of Antibes – 4th international conference on man and animals*, edited by J. Desse and F. Audoin-Rouzeau, 431–7. Juans-les-Pins: Editions APDCA.

Oggins, Robin S. 1980. 'Albertus Magnus on falcons and hawks'. In *Albertus Magnus and the Sciences: Commemorative essays*, edited by J. A. Weisheipl, 441–62. Toronto: Pontifical Institute of Mediaeval Studies.

Oggins, Robin S. 1986. 'Falconry and medieval social status', *Mediaevalia Philosophica Polonorum* 12: 43–55.

Oggins, Robin S. 2004. *The Kings and Their Hawks: Falconry in medieval England*. London: Yale University Press.

Olmos de León, Ricardo M. 2018. 'The care of hunting birds in the late Middle Ages and Renaissance according to the Spanish falconry treatises (1250–1565)'. In *Raptor and Human: Falconry and bird symbolism throughout the millennia on a global scale*, edited by Karl-Heinz Gersmann and Oliver Grimm, 539–55. Kiel: Wachholtz Verlag.

Oppitz-Trotman, Gesine. 2010. 'Birds, beasts and Becket: Falconry and hawking in the lives and miracles of St Thomas Becket', *Studies in Church History* 46: 78–88.

Parry-Jones, Jemima. 2003. *Falconry: Care, captive breeding and conservation*. Newton Abbot: David & Charles.

Payne, John (trans.). 1890. *The Novels of Matteo Bandello, Bishop of Agen*. London: Villon Society.

Petrosillo, Sara. 2023. *Hawking Women: Falconry, gender, and control in medieval literary culture*. Columbus: Ohio State University Press.

Pipe, Alan, Kevin Rielly and C. Ainsley. 2011. 'Animal bone'. In *The Cluniac Priory and Abbey of St Saviour Bermondsey, Surrey: Excavations 1984–95*, edited by Tony Dyson, Mark Samuel, Alison Steele and Susan Wright, 260–3. London: Museum of London Archaeology.

Pluskowski, Aleksander. 2018. 'The medieval wild'. In *The Oxford Handbook of Later Medieval Archaeology in Britain*, edited by Christopher Gerrard and Alejandra Gutiérrez, 141–53. Oxford: Oxford University Press.

Poole, Kristopher. 2018. 'Zooarchaeological evidence for falconry in England, up to AD 1500'. In *Raptor and Human: Falconry and bird symbolism throughout the millennia on a global scale*, edited by Karl-Heinz Gersmann and Oliver Grimm, 1027–53. Kiel: Wachholtz Verlag.

Prummel, Wietske. 1997. 'Evidence of hawking (falconry) from bird and mammal bones', *International Journal of Osteoarchaeology* 7 (4): 333–8.

Resnick, Irven M. and Kenneth F. Kitchell Jr. 2022. *Albert Magnus and the World of Nature*. London: Reaktion Books.

Salisbury, Joyce E. 1996. 'Human animals of medieval fables'. In *Animals in the Middle Ages: A book of essays*, edited by Nona C. Flores, 49–65. New York: Routledge.

Salisbury, Joyce E. 2011. *The Beast Within: Animals in the Middle Ages*. London: Routledge.

Sellet, Frédéric. 1993. 'Chaîne operatoire: The concept and its applications', *Lithic Technology* 18 (1/2): 106–12.

Serjeantson, Dale. 2009a. *Birds: Cambridge manuals in archaeology*. Cambridge: Cambridge University Press.

Serjeantson, Dale. 2009b. 'Food, craft, and status: the Winchester suburbs and defences in a wider context'. In *Food, Craft, and Status in Medieval Winchester*, edited by Dale Serjeantson and H. Rees, 166–83. Winchester: Winchester Museums Service.

Serjeantson, Dale. 2009c. 'Birds: food and a mark of status'. In *Food in Medieval England: Diet and nutrition*, edited by Chris. M. Woolgar, Dale Serjeantson and Tony Waldron, 131–47. Oxford: Oxford University Press.

Serjeantson, Dale. 2010. 'Extinct birds'. In *Extinctions and Invasions: A social history of British fauna*, edited by Terry O'Connor and Naomi Sykes, 146–55. Oxford: Oxbow Books.

Serjeantson, Dale. 2023. *The archaeology of wild birds in Britain and Ireland*. Oxford: Oxbow Books.

Standley, Emma. 2008. 'Ladies hunting: A late medieval decorated mirror case from Shapwick, Somerset', *Antiquaries Journal* 88: 198–206.

Sykes, Naomi. 2009. 'The impact of the Normans on hunting'. In *Food in Medieval England: Diet and nutrition*, edited by Chris M. Woolgar, Dale Serjeantson and Tony Waldron, 162–75. Oxford: Oxford University Press.

Sykes, Naomi. 2010. 'Deer, land, knives and halls: Social change in early medieval England', *Antiquaries Journal* 90: 175–93.

Sykes, Naomi. 2014. *Beastly Questions: Animal answers to archaeological issues*. London: Bloomsbury Academic.

Van den Abeele, Baudouin. 2018. 'Medieval Latin and vernacular treatise on falconry (11th–16th c.): tradition, contents, and historical interest'. In *Raptor and Human: Falconry and bird symbolism throughout the millennia on a global scale*, edited by Karl-Heinz Gersmann and Oliver Grimm, 1271–88. Kiel: Wachholtz Verlag.

Vaughan, Richard. 1982. 'The Arctic in the Middle Ages', *Journal of Medieval History* 8 (4): 313–42.

Wheater, Isabella. 2018. 'Peynte it with aves: Langland's hawks, covetise, and Hugh of Fouilloy's Aviarium'. In *New Medieval Literatures 18,* edited by Laura Ashe, Philip Knox, David Lawton and Wendy Scase, 131–82. Woodbridge: Boydell & Brewer.

Wood, Casey Albert and Florence Marjorie Fyfe (trans.). 1943. *The Art of Falconry, by Frederick II of Hohenstaufen: Being the De Arte Venandi Cum Avibus of Frederick II of Hohenstaufen*. Stanford: Stanford University Press.

Yalden, Derek W. 2002. 'Place-name and archaeological evidence on the recent history of birds in Britain', *Acta Zoologica Cracoviensia* 45 (special issue): 415–29.

Yalden, Derek and Umberto Albarella. 2009. *The History of British Birds*. Oxford: Oxford University Press.

Zeiler, Jørn. 2010. 'Hunting the hunters: owls and birds of prey as part of the falconers' game bag'. In *Birds in Archaeology: Proceedings of the 6th meeting of the ICAZ Bird Working Group in Groningen (23.8–27.8. 2008)*, edited by Wietske Prummel, Jørn Zeiler and Dick C. Brinkhuizen, 163–8. Groningen: Barkhuis.

3
'I live off them, they live off me': exploring the human–flea feeding relationship in the history of flea circuses

Gaia Mortier

Introduction

Humans have been dealing with parasites for thousands of years (Shin and Bianucci 2021). External parasites, often referred to as ectoparasites, have proven to be a persistent challenge for humanity, despite our best efforts to manage them. They have had tens of thousands of years to adapt perfectly to human bodies, their settlements, and their domestic livestock. Alongside lice, fleas are an excellent example of this host–parasite relationship familiar to modern readers. They are small, jumpy, and a never-ending nightmare to any pet owner. Human society prides itself on its problem-solving abilities, yet has struggled to find a permanent solution to the many parasite infestations we continue to face. The history between humans and their fleas is complex, sometimes marked by curiosity and wonder, at other times by disgust and persecution. When viewed through the lens of consumer audiences used to highly engaging and high-production-value digital entertainment services, historical leisure activities seem quaint and foreign by comparison. The popularity of the flea circus as a source of entertainment encapsulates this shift in what audiences are willing to devote their time, money, and attention to. Yet, flea arts were a huge hit in the early twentieth century and have been around since the sixteenth. This chapter will explore a rather atypical human–animal feeding interaction: flea circuses make for an interesting case study of the voluntary feeding of one's own blood to the six-legged performers, and allow for the exploration of the history between humans and their own set of personalised parasites.

What makes a flea?

The word 'flea' is well-recognised, with many being aware that they are a species of tiny animal that feeds externally on a larger animal. Yet, despite their widespread presence and significance, many people would fail to elaborate on what makes a flea. Fleas are nimble insects belonging to a group called the Siphonaptera, derived from the Greek *siphon*, meaning tube, which refers to their highly-specialised, tube-like mouthparts, which allow them to pierce skin efficiently in order to feed on blood, and *-aptera*, meaning wingless, as they have lost their wings through time due to evolutionary pressures (McGavin 2001). Fleas are an excellent example of form meeting function. Where most other insects have their antennae pointing to the sky, the flea has them pointing downwards in order to better navigate the host upon which they reside. Their mouthparts consist of a microscopic tube-like structure that they insert into their host's body through a powerful, muscle-driven mechanism. They also have comb-like bristles on their legs, which allow them to catch onto their host's fur or feathers, thus making themselves harder to remove through grooming. Despite having lost their wings, they are no less agile. Their flattened bodies allow them to navigate forests of hairs and feathers rapidly; strong, enlarged jumping legs allow them to quickly flee from dangerous situations without being caught, as well as jump onto their host once they have moulted into their adult form. The flea's body is covered in spikes and bristles pointing backward, designed to make their removal a challenge even when directly pulled upon. Unlike lice, they do not spend their entire lives on their host: adult fleas lay eggs on the host in order for them to fall off in their immediate surroundings. During their adulthood, fleas are obligate parasites, meaning they have to feed on the blood of their host in their adult form. Their larvae are less host-dependent, feeding on detritus in nests or dens before undergoing a full metamorphosis, which includes a cocoon stage much like that seen in butterflies (Marshall 1981). Fleas can therefore often be found on species living in semi-permanent nests or dens, such as rabbits, chickens, or even humans, as they do not cope well with nomadic lifestyles. The earliest known fossil evidence of fleas can be traced back to the Cretaceous period (125 million years ago), where they are believed to have fed on feathered dinosaurs (Gao et al. 2019). Today, there are over 2,500 species of flea, with the majority feeding on mammals (96 per cent), while the rest feed on birds (4 per cent) (Attenborough 2015; McGavin 2001). As is the case with many parasites, the reason they diversify into numerous species can be explained through the co-evolution between them and their hosts, evolving excellent solutions in order to feed and breed more efficiently in a world of fur and feathers.

Fleas can be particularly difficult to get rid of, as they are quick to reproduce, hard to find and catch, and able to undergo a hibernation period during pupation that can last for months until the conditions are right for them to hatch (Marshall 1981). They are well-known for their appearance on beloved pets, such as cats and dogs. Although largely unfamiliar to the modern reader, the so-called 'human flea' *Pulex irritans* was once a familiar companion across the human race (Whitaker 2007). Despite the name, this species of flea is not particularly host-specific, and is known to feed on a variety of mammals, including domestic pigs, alongside humans (Marshall 1981). Fortunately, they are only vectors for dangerous diseases in extremely rare cases (Miarinjara et al. 2021). Unfortunately, they still cause irritation and discomfort when they decide to make a meal out of you. It is likely that we gained this personalised parasite in the first place from our association with guinea pigs (*Cavia porcellus*), which were first domesticated in South America some seven thousand years ago (Dittmar 2000). *P. irritans* was known to infect guinea pigs, and still does to this day, and is believed to have jumped from them onto us. However, the exact origins of the human flea are still heavily debated due to its complex history in association with its human hosts. Several species of flea have been responsible for the spread of diseases in humans that have led to numerous casualties worldwide. They are often associated with a lack of cleanliness, poor hygiene and unsanitary lifestyles, even though fleas can live in the cleanest of environments. The Romans thought of fleas as animated dirt, specks of dust come to life (Lehane 1969). This complex relationship between humans and fleas has continued throughout history, marked by a mixture of hatred, fear, disdain and, surprisingly, adoration, especially as fleas featured as stars of the show in their own circuses.

History of the flea circus

It comes as a surprise to many that flea circuses were, in fact, real. To understand where the flea circus comes from, and where they have gone, we must find its roots in the sixteenth century – a time when watchmakers and blacksmiths took to creating miniatures. They did so to show off their skills and handiwork, challenging themselves to create the most intricate metalwork designs on a minuscule scale, weighing 'but one grain of gold' (c. 65 mg; Furgurson 2011). This is where the flea, specifically the human flea *P. irritans*, comes in. Attached to the structures by a small chain around the necks, they were selected to demonstrate the lightweight pieces of hardware. This set in motion a series of increasingly extravagant creations for the fleas to pull forward, such as carriages holding 'figures of six horses, a coachman, a dog between his legs, four persons inside, two footmen behind, and a postillion on the fore horse' (Moore 2012) crafted by esteemed watchmaker Sobieski Boverick (1718–1774), or 'a First-rate Man of War, of one hundred and twenty guns, with rigging sails, anchor, and every thing requisite in a three-decker not omitting a numerous crew' (Bertolotto 1835). This delighted audiences and steadily grew in popularity, but the first official record of a flea circus was not until the early 1800s. They reached their highest point in popularity during the early 1900s, where they found a home among the side-shows of most circuses and fairs. The typical flea circus would take place in a small, dimly lit, almost box-like building that would allow crowds to get up close and personal with the six-legged performers.

Louis Bertolotto (1802–1887), an Italian showman living in London at the time, became one of the first and biggest names in the flea circus industry with his 'Extraordinary Exhibition of the Industrious Fleas'. In 1835, Bertolotto went on to write *The History of the Flea with Notes and Observations*, which has passages through which his passion for the small performers become clear. 'The Flea,' he starts, 'when examined by the microscope, affords a very pleasing object … its eyes are very large and beautiful' (Bertolotto 1835, 8). He describes the strength these small insects possess, claiming – from tapping into his circus experience – that they are capable of drawing forward contraptions five hundred times their own body weight. It is clear that Bertolotto was very fond of his fleas. He likens himself to a shepherd knowing his flock, being able to name them without fear of mistaking one for the other, describing some as more stubborn than others. Bertolotto, like many other flea circus professors, was feeding his flea performers with his own blood: 'I have sometimes twenty on my hand, all feeding at once' (Bertolotto 1835, 18). Although some instances of 'blood-for-hire' have been reported, namely an example from the city of Surat, India, where 'some poor fellow, for hire, suffers himself to be tied down upon a bed and the vermin feast upon his body' (Bertolotto 1835), this was not common practice. The flea circus ringmasters, or flea professors, took mostly upon themselves the task of feeding their own livestock, as one does with a hunting dog, or perhaps a cow. This is essentially the same as other acts of animal feeding: humans wish to bond with an animal, or provide some sort of care for it, and thus feed it. One difference is that they provide the nutrition by sacrificing their own health, yet interactions of human–animal feeding often involve a sacrifice of some sorts, albeit mostly monetary. Flea professors chose to closely mimic the flea's 'natural' feeding behaviours, while simultaneously providing enrichment to their star performers, similar to what occurs during public zoo feedings, as discussed in Chapters 8 and 9, with entertainment also playing a role in conservation feeding, as presented in Chapter 11. After every performance, they would stick their flea performers atop their arm, sometimes in front of the audience, not only to offer them a well-deserved meal, but also to provide a shock to the audience. 'I live off them', the professors would reason, 'they live off me' (Lawton 2012).

The act of training an animal is nothing new to the world of circuses, but this was usually done with large animals, such as tigers and lions. Protective legislation and the rising public interest in animal welfare have led to animal-based circus acts being generally frowned upon: at least within the UK, tigers are no longer allowed to jump through hoops and elephants cannot be made to balance upon a stool many times too small for their size (Rizzolo 2015; Wild Animals in Circuses Act 2019). But go back a few centuries and these animal acts were all the rage – the outcries of a select few organisations usually falling on deaf ears. Plenty of flea professors claimed to have trained their performers, whereas others deny it required any training at all (Berquam 1977). That is not to say it required no skill – quite the opposite, in fact: a steady hand and patience were required to tie the miniature chains and wires, the width of a single hair, around the tiny necks of fleas. Those who trained their fleas did so using a rather harsh method. The fleas were placed in a tube with sides containing a glue-like substance such as Vaseline, in which the fleas would become trapped if they tried to jump. This was all aimed at teaching the little animals that escape was no longer an option; they took the act of jumping away from the fleas. Instead, the fleas would learn to hold still and do as they were told – at least, that is what the flea trainers claimed to achieve (Lehane 1969).

A question immediately arises: why human fleas? Why fleas at all? The first reason is the most straightforward: ease of access. The human flea was extremely common globally before the twentieth century. It was therefore little effort to acquire new performers once the old set had run out of steam, as human fleas typically live for no more than three months: simply ask a friend or family member to have a quick brush of their clothing (McGavin 2001). To understand the second reason as to why fleas – specifically human fleas – we must turn to the biology of the flea, as it is this that allows them to be the greatest of showmen. Despite flea professors' best efforts to prevent them doing so, fleas are known for their ability to jump great distances. Having evolved not only to access hosts in order to feed, but also to escape from predation in a matter of milliseconds, fleas are capable of jumping eight times their body size in only one-thousandth (0.001) of a second (Furgurson 2011). This had puzzled scientists for decades, as their muscles alone did not seem capable of exerting such force. The answer is a protein called resilin, which is stored in the flea's strong hind legs and is able to store large amounts of energy (Lyons et al. 2011). When this is released, the flea is catapulted into the air. It is this same strength that makes them such fantastic circus assets, capable of pulling, pushing, and moving structures much larger than themselves. As for human fleas, flea professors claimed that their hind legs were much more powerful than other common flea species, such as the cat or dog flea. However, this has since been questioned, as the suggested size difference between species is not consistent, and depends heavily on the individual flea.

Even though the biology explains the 'why' of flea circuses, it does not clearly answer why such a circus was appealing in the first place. That is, were people not disgusted, or even scared, of these tiny blood-sucking insects? Aside from the irritation and infestations they could cause, they are now well-known as vectors of disease, notably the bubonic plague or Black Death. Yet it is important to keep in mind that it was not until the late-nineteenth century that fleas were held responsible for the spread of this plague, when in 1898 biologist and chief medical officer Paul-Louis Simond (1858–1947) verified the presence of the bacterium responsible for the plague (*Yersinia pestis*) within oriental rat fleas (*Xenopsylla cheopis*, see Miarinjara et al. 2021; Mollaret 1999). Despite this scientific discovery, fleas remained one of the most sought-after animal performers of the early twentieth century. One could argue that since the human fleas used in the circuses are a different species from the oriental rat flea, they are therefore not as inherently dangerous and not associated with the spread of the plague – but it cannot be confidently stated that the public would be equipped to identify or tell apart the two species, especially in a dimly lit room from a distance, something probably impossible even for an experienced entomologist (Miarinjara et al. 2021). Perhaps the danger was part of the excitement; with controlling this 'dangerous', tiny animal, came a rush of dominance – the sense that we as humans were so clearly able to dominate and control a species that would otherwise seek to prey upon us. Whatever the case, during performances, fleas were chained or fastened to prevent their escape, and subsequent infection of the audience.

The demise of the flea circus is heavily tied to the demise of the human flea. This human-blood-thirsty insect, which 'fattens at the expense of the human species' (Bertolotto 1835, 11), was believed to be much stronger than its relatives who feast on cats and dogs, and until the mid-twentieth century was readily available. Therefore, the rapid post-Second-World-War decline in the population of *P. irritans* had a substantial impact on the flea circus economy. The decline is largely linked to the widespread availability of the vacuum cleaner, and subsequent increase in dry hygiene practices, which are detrimental to flea eggs. With the downfall of *real* flea circuses came the rise of the *fake* flea circus. Flea professors, not wanting to lose out on their clientele, replaced the real fleas with magnets, mechanisms, or in some cases even a combination of both with a dead flea (likely to be from a cat or dog flea) glued on (Berquam 1977; Furgurson 2011). Signs would read 'Don't Be Sceptical; Seeing is Believing', announcements would be shared that they would be 'Positiflea an All-live Show' – but most of these were false promises. The replacement of human fleas with a different species was proposed, yet this suggestion did not seem to garner much popularity. As discussed earlier, cat and dog fleas were considered not to have sufficient strength. Perhaps the fact that they were human-feeding fleas added to the allure: taming that which feeds on us in such a military manner. The success of the no-flea flea circus likely contributed to the complete removal of real fleas from the circuses, with a shift in focus to the showmanship of the ringleader. Nowadays, one might be hard-pressed to locate a surviving flea circus show, though a passionate few are keen on keeping the tradition alive. One of the last remaining flea circuses travels with Oktoberfest, which takes place yearly in Munich, Germany; but the answer to the question whether or not they are using real fleas is the same: seeing is believing.

The history of fleas and their humans

Though the cultural impact of flea circuses cannot be underestimated, they were merely a part of the entertainment fleas provided throughout history (Lehane 1969). The art of dressed fleas, or in Spanish *pulgas vestidas*, originates from nineteenth-century Latin America, around the Mexican city of Guanajuato to be precise, where nuns took to creating miniature settings to exercise their minds. They sewed tiny clothes, not to actually dress the fleas like a doll, but to stage the heads of fleas atop the costumes to create the illusion of dressed fleas. This tradition later on became a popular tourist keepsake well into the early 1900s, with the most sought-after specimens being those dressed up as a wedding pair, or typical mariachi bands – complete with instruments. Octavio Paz, winner of the 1990 Nobel Prize in Literature, said the artform was 'difficult, exquisite and useless' (Harding n.d.). Similar to the art of the flea circus, this hobby also saw a steep decline in the early-twentieth century, leaving original dressed fleas to become a highly sought-after and rare collectible.

Before the seventeenth century, there were many theories as to what brought small animals, such as insects, into existence. That is, if they were even considered animals. It was not until the invention of the microscope that many accepted that fleas sexually reproduce, and did not birth from the dust collecting on the windowsill, an idea expressed in Edward Topsell's *The History of Four-Footed Beasts and Serpents*: 'the Latin word *Pulex*, comes from *Pulvis*: dust, or the son of dust' (Topsell 1607, 1127). The ability to observe insects closely led to an increased understanding of their form and function. In 1610 Galileo used a rudimentary microscope and discovered that insects have incredibly complex eyes. Antonie Philips van Leeuwenhoek (1632–1723) was one of the first to dedicate his craft to the microscopic world. He used microscopes of his own design to observe what he described as *diertjes* (Dutch for small animals), what we now know to be microbes. He not only described a flea in detail, but also the mite that lived on its pupa (Lehane 1969). Furthermore, he concluded that the flea was in fact not 'produced from corruption, but in the ordinary way of generation'. Robert Hooke (1635–1703) was a scientist and author of *Micrographia*, one of the greatest works of its time. The pioneering book featured stunning images of minuscule structures, animals, and plant life as Hooke observed them through a microscope.

One of the largest and well-kept flea collections is the Rothschild collection, which currently resides in the Natural History Museum at Tring. Their collection contains more than a quarter of a million specimens, representing over 75 per cent of all known species and subspecies of flea as described today, from all over the world. Charles Rothschild (1877–1923) was a British banker turned entomologist with a particular passion for fleas, describing over five hundred new species during his lifetime. He suffered from severe brain inflammation for most of his life, and tragically took his own life when he was only 46 years old. His brother Lionel Walter Rothschild (1868–1937) was also a cunning zoologist, opening his private museum in 1892 to store his large natural history collection, which has since become the Walter Rothschild Zoological Museum section of the Natural History Museum at Tring.

Charles' daughter Miriam Rothschild (1908–2005) was keen to follow in her father's footsteps. She went on to become a pioneering woman within entomology and the leading authority on fleas; being the first person to work out how exactly fleas are able to jump with such strength. She furthermore established the link between the flea's reproduction and the host's hormonal cycles – as seen in rabbits and rabbit fleas (*Spilopsyllus cuniculi*); the fleas are able to detect changes in certain hormones within the female rabbit's blood, which signify that she is about to give birth, which in turn triggers the fleas to reach sexual maturity (Rothschild and Clay 1952). She contributed hugely to the field of siphonapteran studies, publishing numerous well-received works, among which was the popular book *Fleas, Flukes and Cuckoos*, which she wrote together with fellow-entomologist Theresa Clay (1911–1995), who specialised in parasitic lice.

Humans and their domestic livestock have been closely and intimately linked to their own set of personalised parasites for thousands of years; mainly – as emphasised – fleas, but also species of lice, mites, and ticks (Marshall 1981). Lice are small, flat and wingless animals belonging to the insect order *Phthiraptera*. They spend their entire life cycle on their host and cannot survive for long without them – they only jump ship once the host itself has died or an opportunity for dispersal presents itself. Mites are arachnids that form the group *Acari*, along with ticks. Ticks are most well-known for their contribution to the spread of Lyme disease in humans, caused by the bacterium *Borrelia burgdorferi*, whereas several species of mite live a relatively harmless life inside our pores (Krantz and Walter 2009). In this context, fleas are unique due to their ability to survive for longer periods off-host, and to freely pick and choose which one to feed and breed on (McGavin 2001). That is not to say that fleas have no preference; they usually limit themselves to a small range of species within their immediate proximity. In the context of domestication this has proven to be the perfect playground for the spread of fleas from livestock to humans. As mentioned, the human flea (*P. irritans*) has been known to infest guinea pigs, dogs, and humans alike – creating the perfect bridge from livestock to human hosts (Dittmar et al. 2003b). Similar instances can be observed in modern contexts where human owners get bitten by the fleas that live on their pets, most commonly cats or dogs. Until the eighteenth century ladies would wear a garment called a 'flea fur', which consisted of the pelt of an animal designed to attract fleas away from the person themselves by offering a false feeding opportunity (Bain 2004).

Fleas are infamous for their now heavily debated role in the spread of the Black Death, one of the deadliest pandemics throughout human history (Miarinjara et al. 2021). This fourteenth-century plague ultimately led to the death of more than two hundred million people and is caused by the bacterium *Yersinia pestis*. It has long been thought that this disease was primarily transmitted through the bites of rat fleas (*X. cheopis*) that were feeding on infected rats. The fleas themselves would not get ill as the bacteria multiplied within its gut. After the rats succumbed to their illness, the fleas would be on the lookout for a new host to feed on – which in many cases ended up being the uncovered ankles of humans. Fleas formed the perfect vector, as when they pierced the skin in order to feed on the human's blood, the bacterium would find its way into the bloodstream before multiplying and causing a range of symptoms that included fever and chills. The most notable symptom was the appearance of painful black swellings called buboes, often located in the groin, armpit or neck. This is what ultimately led to the disease being titled the 'bubonic plague' (Dean et al. 2018). In recent years, several studies have put forward alternate theories about the spread of this historic plague pandemic, such as it being airborne, or caused by contaminated food or water sources. Although these factors may have played a small role at the time, there is an overwhelming body of evidence that supports the vital role of fleas in spreading the disease in modern cases. There have been several instances of fleas being used as biological weapons in order to spread the plague through enemy territory. Historically, this form of biological warfare was likely applied to sieges: bubonic plague-ridden bodies covered in fleas would be catapulted or thrown over the walls of besieged cities to rapidly spread the disease (Wheelis 2002). The most recent instance of flea-based warfare occurred during the Second World War when the Japanese army aimed to weaken Chinese defences (Dennis 2009; Stewart 2011). The Japanese Army's Unit 731 developed and executed the attacks that were responsible for significant plague outbreaks during the early 1940s, notably those in Ningbo and Changde. The disease would not remain contained to the target population, however, and eventually led to thousands of deaths, mainly of civilians; due to the secretive nature of the attacks, the exact number of casualties is difficult to estimate.

The close association and co-evolution between parasites and their hosts mean they can be used as an opportunity to learn about things that would otherwise be out of our reach, especially within ancient archaeological contexts. Here, human remains often consist of only bones, hair, teeth and nails. In the case of external parasites, their toughened exoskeleton often allows them to be preserved exceptionally well, especially in the case of mummified remains or those found in anaerobic conditions in waterlogged areas. External parasites have been used within the field to imply patterns of human migration, the spread of diseases, and environmental factors such as temperature and precipitation (Raoult et al. 2008). Fleas have been recovered from archaeological excavations in Egypt dating back to the reign of Tutankhamun (c. 1350–1323 BCE). As juvenile fleas do not live on their host, and instead in their surroundings, they are most commonly found within well-preserved clothing or furniture, or areas where large numbers of livestock were kept. This is especially true for flea eggs and cocoons, as they are able to lie dormant for an extended period of time, and in some cases being killed and preserved by the mummification of their hosts before they were able to hatch. Hundreds of adult fleas have been uncovered within the mummified remains of domesticated animals from South America, dating back to pre-Columbian times roughly 1,100 years ago (Dittmar et al. 2003a). They have furthermore been recovered from settlements in the Netherlands (c. 900–700 years ago), Greenland (c. 1,100–700 years ago), and Egypt (c. 3,550 years ago) (Panagiotakopulu 2001; Sadler 1990; Schelvis 1997). As the availability of laboratory-based techniques increased, a small number of archaeological studies successfully attempted methods of biochemical analysis from uncovered external parasites, such as the extraction of ancient DNA (aDNA). A commonly used technique in archaeology is that of stable isotope analysis, which is used to explore past diet and mobility using biological tissues, such as hair, teeth, and bone. Despite this, this technique has not yet been applied to parasite remains uncovered together with their hosts. It is safe to say there is plenty left to explore within this field of biochemical-based archaeoparasitology. The use of these techniques on parasites is especially promising as they would be closely linked to their hosts, which in some cases have gone extinct, such as the parasitic fly *Cobboldia russanovi* that used to live on mammoths (Cascardo et al. 2021; Grunin 1973). This could not only provide a unique range of information if the blood meal is preserved inside the tough exoskeleton, but could also circumvent the need for invasive sampling of the host.

The field of forensics also employs many of the techniques used in archaeology, in order to find out details about a victim's location, health conditions and, in some cases, association with specific animal species. Several species of flea are commonly associated with humans, and therefore have been of use in forensic investigations; the cat flea (*Ctenocephalides felis*), dog flea (*Ctenocephalides canis*), human flea (*P. irritans*), and jigger flea (*Tunga penetrans*), the latter being unique among fleas as the females live and feed subdermally. As fleas found on the scene often still contain the blood from their hosts, they have in some cases been analysed for DNA in order to place a person within the crime scene. Large infestations are used as indicators of neglect or poor hygiene, yet despite their close association with humans, they are not ideal species for forensic investigations. Fleas are quick to jump ship once their host has died and therefore are not expected to stick around for long after death. Furthermore, they will not feed on a deceased body and therefore cannot be used to indicate a time of death, unlike maggots or beetles (Smith 1986).

The importance of parasites

Parasites, as a whole, are severely underappreciated when it comes to ecological conservation, despite forming just under half of all described species (Dobson et al. 2008). They are often forgotten when people talk about which animals keep an ecosystem from falling apart, despite their huge importance as both predator and prey. They form two trophic levels, as the food source for numerous species, but also as a critical predator keeping population levels in check. Without them, as with many other insect species, entire ecosystems would collapse. Almost every multicellular animal on this planet has its own range of parasites. Some are mutualists and provide vital services to their host in a win-win situational way of living. Some are incredibly influential on their surroundings, acting as ecological keystone species and moulding entire ecosystems; others exist just to feed themselves on another, reproduce, and die. Parasites additionally form an important evolutionary pressure, forcing their hosts to adapt and overcome diseases: a filter through which the weaker individuals will die and not pass on their genes. They alter community structures as they target specific host types in a form of parasite-mediated competition. An excellent example of this is the grey squirrel (*Sciurus carolinensis*), an animal considered invasive throughout the UK, which brought with it a parasite that the native red squirrel (*S. vulgaris*) was not adapted to. This led to drastic declines in red squirrel populations, allowing the grey squirrel to take over their territory (Wood and Johnson 2015).

Fleas may be a nuisance to our pets, and to ourselves, but the ways we eradicate them from our households are causing much more damage on a large scale (Perkins and Goulson 2023). The most commonly used treatments contain imidacloprid and fipronil – two substances that have long since been banned from agricultural use for being too damaging to non-target species. In fact, one month's worth of flea treatment for a large dog is enough to kill millions of insects, and these compounds are now finding their way into our freshwater ecosystems as pollutants. Globally, many species of flea are rapidly declining, alongside thousands of other species as the twenty-first-century biodiversity crisis goes on. Yet we are not sad about the population decline of the human flea, our own personal loss, in the way we are sad about the loss of the dodo. It is misleading to say that it is not in our nature to care for those animals that mean us harm. Any cat owner will proudly retell the story of the latest scratch they have obtained while trying to show affection to their pet, and we do not blame the tiger for lashing out at us if we stand too close to their cage – it is in their nature. If humans pick and choose what nature is allowed to continue, the world will contain nothing more than 'functional' biodiversity. The cow can stay, it brings us milk; the bee can live on, for it provides us with honey and crops. But what about the frog, who relies on the mosquito to feed, or the beetle, who devours fleas for almost every meal (Lehane 1969)? Perhaps it was this that prompted the flea professors to feed their performers with their own blood; a simple exchange of services. But from Bertolotto's writing it seems clear that there was also an element of care involved, akin to a farmer tending to his flock of sheep. Whatever the case may be, the once one-sided battle between humans and their fleas seems to be reaching its conclusion, yet it remains unclear if it can be considered a victory.

Conclusion

The history between humans and the nimble, blood-sucking insects named fleas underline the complicated nature of human–animal relationships. Their incredible features are the result of thousands of years of adaptations to perfectly feed and breed on their preferred hosts. Fleas have bound themselves to humans and their domestic livestock through such adaptations. Aside from irritation, they are held responsible for the deaths of over two hundred million people due to their efficiency as a vector for disease. Yet our history with these parasites is not all negative. They have provided entertainment as circus performers, provide irreplaceable ecological services and form an integral part of any ecosystem they are found in (which is most of them), and due to their close association with humanity, they provide exciting opportunities for studies of our past and future. All in all, the relationship between humans and animals such as fleas remains turbulent; be it love or hate, one thing is for certain: we are intimately linked.

References

Attenborough, David. 2015. *Natural Curiosities*, Series 3, Episode 1: 'Impossible feats' (TV documentary). BBC2 England.
Bain, Allison. 2004. 'Irritating intimates: The archaeoentomology of lice, fleas, and bedbugs', *Northeast Historical Archaeology* 33: 81–90.
Berquam, David. 1977. 'The exchange: Flea circus', *Renaissance Quarterly* 16 (4): 309–11.
Bertolotto, Louis. 1835. *The History of the Flea: With notes and observations*. London: Crozier.
Cascardo, P., E. Pucu and D. Leles. 2021. 'Review of parasites found in extinct animals: What can be revealed', *Journal of Parasitology* 107 (2): 275–83.
Dean, Katharine R., Fabienne Krauer, Lars Walløe, Ole Christian Lingjærde, Barbara Bramanti, Nils C. Stenseth and Boris V. Schmid. 2018. 'Human ectoparasites and the spread of plague in Europe during the second pandemic', *Proceedings of the National Academy of Sciences* 115 (6): 1304–9.
Dennis, David T. 2009. 'Plague as a biological weapon'. In *Bioterrorism and Infectious Agents: A new dilemma for the 21st century*, edited by I. W. Fong and Kenneth Alibek, 37–70. New York: Springer.
Dittmar, Katharina. 2000. 'Evaluation of ectoparasites on the guinea pig mummies of El Yaral and Moquegua valley, in southern Peru', *Chungara: Revista de antropología Chilena* 32 (1): 123–5.
Dittmar, Katharina, U. Mammat, M. Whiting, Sonia Guillén and Karl J. Reinhard. 2003a. 'Techniques of DNA-studies on prehispanic ectoparasites (*Pulex* sp., Pulicidae, Siphonaptera) from animal mummies of the Chiribaya culture, southern Peru', *Papers in Natural Resources* 56 (98): 53–8.
Dittmar de la Cruz, Katharina, Regine Ribbeck and Arwid Daugschies. 2003b. 'Palaeoparasitological analysis of guinea pig mummies of the Chiribaya Culture, Moquegua valley, Peru', *Berliner und Münchener Tierärztliche Wochenschrift* 116 (1/2): 45–9.
Dobson, Andy, Kevin D. Lafferty, Armand M. Kuris, Ryan F. Hechinger and Walter Jetz. 2008. 'Homage to Linnaeus: How many parasites? How many hosts?', *Proceedings of the National Academy of Sciences* 105 (1): 11482–9.
Furgurson, Ernest. 2011. 'A speck of showmanship: Is that a *Pulex irritans* pulling that carriage, or is someone just pulling our leg?', *American Scholar* 80 (3): 92–7.
Gao, Taiping, Xiangchu Yin, Chungkun Shih, Alexandr P. Rasnitsyn, Xing Xu, Sha Chen, Chen Wang and Dong Ren. 2019. 'New insects feeding on dinosaur feathers in mid-Cretaceous amber', *Nature Communications* 10 (1): 1–7.
Grunin, K. Y. 1973. 'The first finding of the stomach bot-fly larvae of the mammoth: *Cobboldia* (*Mamontia*, subgen. n.) *russanovi*, sp. nov. (Diptera, Gasterophilidae)', *Entomological Review* 52: 228–31.
Harding, Deborah. n.d. 'Dressing fleas'. Carnegie Museum of Natural History. Accessed 29 May 2023. https://carnegiemnh.org/dressing-fleas/.
Krantz, G. W. and David Evans Walter. 2009. *A Manual of Acarology*. Lubbock: Texas Tech University Press.
Lawton, Graham. 2012. 'Fleadom or death: Reviving the art of the flea circus', *New Scientist* 216 (2897): 53–5.
Lehane, Brendan. 1969. *The Compleat Flea*. London: John Murray.
Lyons, Russell E., Darren C. C. Wong, Misook Kim, Nicolas Lekieffre, Mickey G. Huson, Tony Vuocolo, David J. Merritt, Kate M. Nairn, Daniel M. Dudek, Michelle L. Colgrave and Christopher M. Elvin. 2011. 'Molecular and functional characterisation of resilin across three insect orders', *Insect Biochemistry and Molecular Biology* 41: 881–90.
Marshall, Adrian. 1981. *The Ecology of Ectoparasitic Insects*. London: Academic Press.
McGavin, George C. 2001. *Essential Entomology: An order-by-order introduction*. Oxford: Oxford University Press.

Miarinjara, Adélaïde, David M. Bland, James R. Belthoff and B. Joseph Hinnebusch. 2021. 'Poor vector competence of the human flea, *Pulex irritans*, to transmit *Yersinia pestis*', *Parasites and Vectors* 14 (1): 317.

Mollaret, Henri H. 1999. 'The discovery by Paul-Louis Simond of the role of the flea in the transmission of the plague', *Bulletin de la Société de Pathologie Exotique* 92 (2): 383–7.

Moore, Keith. 2012. 'The ghost of a flea'. Royal Society. Accessed 5 June 2023. https://royalsociety.org/blog/2012/10/the-ghost-of-a-flea/.

Panagiotakopulu, Eva. 2001. 'Fleas from pharaonic Amarna', *Antiquity* 75 (289): 499–500.

Perkins, Rosemary and Dave Goulson. 2023. 'To flea or not to flea: Survey of UK companion animal ectoparasiticide usage and activities affecting pathways to the environment', *PeerJ* 11: e15561.

Raoult, Didier, David L. Reed, Katharina Dittmar, Jeremy J. Kirchman, Jean-Marc Rolain, Sonia Guillen and Jessica E. Light. 2008. 'Molecular identification of lice from pre-Columbian mummies', *Journal of Infectious Diseases* 197 (4): 535–43.

Rizzolo, Jessica Bell. 2015. '"There is no wild": Conservation and circus discourse', *Society and Animals* 23 (5): 462–83.

Rothschild, Miriam and Theresa Clay. 1952. *Fleas, Flukes and Cuckoos: A study of bird parasites*. London: Collins.

Sadler, J. P. 1990. 'Records of ectoparasites on humans and sheep from Viking-age deposits in the former western settlement of Greenland', *Journal of Medical Entomology* 27 (4): 628–31.

Schelvis, Jaap. 1997. 'Caught between the teeth: A review of Dutch finds of archaeological remains of ectoparasites in combs', *Proceedings of the Experimental and Applied Entomology Section of the Netherlands Entomological Society* 5 (4): 131–2.

Shin, Dong Hoon and Raffaella Bianucci. 2021. *The Handbook of Mummy Studies: New frontiers in scientific and cultural perspectives*. Singapore: Springer Verlag.

Smith, Kenneth G. V. 1986. *A Manual of Forensic Entomology*. Ithaca, NY: Cornell University Press.

Stewart, Amy. 2011. *Wicked Bugs: The louse that conquered Napoleon's army and other diabolical insects*. London: Timber Press.

Topsell, Edward. 1607. *The History of Four-Footed Beasts and Serpents*. London: Printed by William Iaggard.

Whitaker, Amoret. 2007. *Fleas (Siphonaptera)*. (*Handbooks for the Identification of British Insects*, Volume 1, Part 16). London: Royal Entomological Society.

Wheelis, Mark. 2002. 'Biological warfare at the 1346 siege of Caffa', *Emerging Infectious Diseases* 8 (9): 971–5.

Wild Animals in Circuses Act. 2019. 'Prohibition on use of wild animals in travelling circuses in England (c.24)'. Accessed 22 May 2024. https://www.legislation.gov.uk/ukpga/2019/24/enacted.

Wood, Chelsea L. and Pieter T. J. Johnson. 2015. 'A world without parasites: Exploring the hidden ecology of infection', *Frontiers in Ecology and the Environment* 13 (8): 425–34.

4
Human–raptor relationships in urban spaces: the history of red kites (*Milvus milvus*) and human food in Britain

Juliette Waterman

Introduction

Red kites are a common sight in many areas of the UK today, but they have a complex history, with what seems to be alternating periods of legal protection and persecution. Through this history of turbulent interaction with humans, they have coexisted with people and our settlements, drawn by the ample sources of food we introduce into our landscapes. It is this relationship with food that seems to define our interactions with the red kite. They are commensal animals – those that dine 'at our table' and benefit from association with humans in the urban ecosystems we govern – a category that includes many other familiar faces such as rats, foxes, and pigeons (O'Connor 2013). The red kite may seem an unlikely inclusion among such company to any who live in areas where they are rare or absent, but their dietary niche is equally well-suited: they are a predatory bird, but also one that scavenges, and they are generalists who take advantage of a wide range of food items if offered. This provides them with lots of opportunities to interact with humans and the food resources associated with us. The relationship produced by this shared engagement with food has taken many forms, including the extraction of waste food from urban spaces, conflicts over 'theft' of food, the use of 'dangerous' food as a means of environmental control, and management of food access for conservation. More generally, the foodstuffs that characterise the human–commensal relationship can be categorised broadly into foods that humans deem acceptable for consumption by our animal neighbours, and those that are not – but the precise nature and reasoning behind these porous categories varies enormously over time. The history of the entanglement between humans and red kites is described along with the changing role of food as a mediator in this multifaceted relationship.

Urban ecologies: the role of 'waste' in supporting commensals

Many scavenging species benefited from the urban ecosystems of medieval and early modern Britain, where food was plentiful in the form of waste and supported a diverse suite of commensals (O'Connor 2013). Multiple lines of evidence from environmental history and archaeology suggest that red kites were among those present in these cities, indicating that they were able to adapt their foraging to this food-rich environment. O'Connor (2000) emphasises the value of organic waste deposits in medieval settlements in supporting diverse urban food webs, with red kites as a key example. Further discussion on the role of scavenging raptors as urban commensals is provided by Mulkeen and O'Connor (1997), with many examples of kite remains excavated from medieval urban sites. Other authors also review the presence of birds in the archaeological record throughout British history, such as the regional reviews of archaeological material of British sites published by Historic England (Albarella 2019; Holmes 2017), Yalden and Albarella (2009) and Serjeantson (2023). A synthesis of this material as well as the faunal data accessible through the Archaeology Data Service indicates a widespread distribution of kites across the UK; their past abundance in England specifically may however reflect bias in where archaeological work is conducted and published rather than indicating historical absence elsewhere. Although this chapter begins with the medieval period, as it is particularly rich in source material, like many commensals the kite may have associated with humans from the earliest period of human settlements: they are found from the very earliest phases of British prehistory (Yalden and Albarella 2009, 36), and their relationship with humans and food acquisition may have begun in the UK as they profited from the forest clearances undertaken by Neolithic farmers for crops and pastureland (Lovegrove 1990, 28). In medieval settlements especially, however, the presence of red kites is suggested by the existence of a suitable niche for them as scavengers, which is confirmed by the presence of skeletal remains. They are primarily found in deposits of waste and refuse such as middens or cesspits, and typically represented by a very small number of elements rather than partial skeletons, with no special burials of whole skeletons (unlike those seen for falconry birds, see Chapter 2, and corvids, see Chapter 5, which may reflect a higher degree of cultural importance or ritual engagement with these taxa). The archaeological evidence therefore indicates that they were present at human settlements, but not treated with any particular reverence or as having any particular value.

The archaeological evidence is in keeping with contemporary texts suggesting that red kites were familiar to urban dwellers, such as the Venetian ambassador to England who wrote in c. 1500 that kites 'became so tame as to mingle with the passengers, and take their prey in the midst of the greatest crowds', and 'so tame, that they often take out of the hands of little children, the bread smeared with butter, in the Flemish fashion, given to them by their mothers' (translated in Sneyd 1847, 11). The same ambassador suggests they were valued to the point of being legally protected for their scavenging role – an assertion best understood in the context of the medieval city and the part a commensal could have played in urban life.

The role of red kites in towns becomes clearer in the context of medieval urban sanitation issues, which were both a source of food for commensal animals and also a source of pollution for human residents. Sabine (1937) was first to review measures taken to maintain cleanliness in medieval cities, with their complexity and sophistication indicating the scale of waste management issues but also challenging the idea that they were passively tolerated. Following this, the myriad works of Jørgensen on British cities (especially 2010a, 2010b, 2014, 2020) collate a diverse range of medieval sources on sanitation, such as legislation to tackle waste, centrally organised refuse collection services, fines issued to offenders, and complaints made about non-compliance. From these, a broad and nuanced picture of the practical side of refuse management can be built, grounded in a desire to promote health and wellbeing by removing unpleasant and dangerous sensory experiences (Jørgensen 2021). Within this, smell and its perceived pathogenic properties through the miasmic theory of disease spread was especially crucial (Jørgensen 2013). This litter that so offended city dwellers in turn provided rich pickings for opportunistic scavengers – with butchery waste being especially beneficial in the context of carnivorous scavengers such as the red kite. Such scavengers, by removing this abundant disease-causing waste from the sensory field of medieval citizens, might therefore have been considered valuable. Fortunately for the red kite, butchers and their trade by-products were, in the view of Ciecieznski (2013), the most heavily legislated and reproached source of urban pollution, and Sabine (1933) reviewed the management of butchery waste specifically, to follow his previous work on medieval city cleanliness, suggesting a grand scale and therefore abundant food resource. Carr (2008) reviews measures that address problems caused by butchery waste from across urban England and finds regulations governing its disposal from nearly all towns: these generally restrict where animals could be slaughtered and their offal deposited, and occasionally prescribe areas to dump waste to keep it away from streets and places of habitation. Through specific legislation tackling problem areas, a detailed picture of the urban ecosystem and its potential sources of edible waste emerges. Key concentrations include gutters in between episodes of adequate rainfall, in rivers, which sometimes became blocked from excess waste, and in abandoned plots of land, as well as in officially designated waste pits (Jørgensen 2010a). The image arises from these analyses of medieval civic sources of a careful balancing act between waste generation and waste management, with many potential sources of food for scavengers, but also a concern for minimising or distancing many of these pockets from the city's inhabitants. The extensive scholarship on medieval waste disposal reveals the medieval city as an urban ecosystem with abundant resources available to the streetwise scavenger, a niche made more attractive if they were tolerated and encouraged by the humans sharing these urban habitats, creating a mutually profitable arrangement.

A shared interest could have led to harmony, with medieval town dwellers desiring the removal of waste and scavengers only too happy to oblige if allowed to occupy urban spaces – but does this mean the symbiotic relationship was recognised at the time? Many wildlife historians (such as Gurney 1920; Lovegrove 2007; Cobham 2014) repeat the belief that red kites were legally protected for this reason, and it certainly seems tidy: legislative security for medieval kites would neatly bookend the intervening centuries of persecution – as they now are protected by law under the Wildlife and Countryside Act of 1981. However, Raye (2021) disputes the existence of such a law, for which there is no statutory evidence, and instead considers it to be a fictional theme generated by foreign writers keen to emphasise the relative barbarism of English cities. As well as the Venetian ambassador discussed above, Raye (2021) cites several foreign visitors to London who refer to red kites being numerous and protected by law. The Bohemian squire Schaseck, who visited London with his master Baron Leo of Rozmital between 1465 and 1467, recounted 'I have never seen so many kites as I saw there. It is a capital offence to harm them' (translated in

Letts 1957, 51). Whether the law explicitly protected them or not, there does seem to be a degree of cultural value placed on the red kite for its commensal role. Furthermore, the potential fiction of legal protection is repeated by diverse sources, and may have percolated into public belief in a way that amounted to the same end result, resulting in kites being protected from harm. This belief also made its way into later natural history texts compiled from the work of earlier authors, solidifying the perception that kites did valuable – if unpleasant – work in medieval cities. An example of this is the French naturalist Pierre Belon's 1555 *L'histoire de la nature des oyseaux*, which affirmed that kites in England were protected from violence thanks to their removal of 'filth' from streets and rivers (Belon 1555, 131). This theme also features in the seventeenth-century work of Thames waterman-poet John Taylor who praises the 'good offices' of the red kite 'in devouring and carrying away our Garbage and noysome excrements, which they live by: and if they were not our voluntarie Scavengers, we should be much annoyed with contagious savors of these corrupted offals'; he compared them to the brothel madam who serves as the 'wheelbarrow for the close conveyance of man's luxurious nastinesse' (Taylor 1630, 99). The ignoble scavenging of the kite was tolerable, even valuable, when only waste foodstuffs were consumed – those that had already been removed from circulation as human food. Urban ecosystems, however, also offered plenty of other food resources that were not sanctioned for scavengers, as they were instead earmarked for human consumption.

Food competition and vermin

Medieval and early modern kites may have been somewhat valued for their role in urban sanitation, but contemporary texts can be quite damning of their feeding habits when they are seen to encroach on those foods classed as 'for human consumption'. A key source for this theme is the medieval bestiary. Their encyclopedic format of animals shares similarities with later natural history texts, but bestiaries are very different in aim: although they catalogue the natural world, their function is a moral message rather than pursuit of scientific knowledge. This message can be crucial for understanding medieval perceptions of animals and relationships with nature. Typically bestiaries reproduce the same text to accompany their images of birds (see Figure 4.1): *De avibus* (On Birds), written by Hugh of Fouilloy. In it, the red kite is used as a moral lesson against laziness and greed, and it is the theft of human food that earns

Figure 4.1 Detail from the illuminated folio 46v of the *Aberdeen Bestiary* MS. 24, showing a red kite. *Source*: Aberdeen University Library Special Collections, https://www.abdn.ac.uk/bestiary/ms24/f46v (CC BY 4.0).

it such harsh condemnation – specifically hovering around kitchens and meat-markets to get an easy meal. The reader is warned against similar low temptations, and to avoid being likewise 'very concerned about their stomachs' (Hugh of Fouilloy, in Clark 1992, 207). The kite's habit of snatching young domestic fowl as easy prey rather than hunting wild birds is also compared to the negative influence hedonistic people have on the young and naïve. For this comparison to be relevant to the medieval reader, this format of negative interaction must have been recognisable, in the same way that other, more neutral medieval sources indicate a proximity between kites and town-dwellers. In the bestiary's representation, the red kite is not a valued street cleaner but instead an enemy to orderly human life, and its theft of 'forbidden' foods transforms it into a piece of 'wildness' resisting control.

By the sixteenth century this concern over food competition had developed into a formal legal framework, with the 1566 Acte for Preservation of Grayne (Elizabeth an.8; cap.XV). This law permitted churchwardens to issue bounties for the killing of any of the targeted 'vermin' animals perceived to be a threat to human food sources (Jones 1972). Any person given permission by a landowner could – and was encouraged to – destroy 'vermin' by any 'reasonable Devyse' (guns and crossbows excluded) and present the heads or eggs to their local churchwarden for payment. Table 4.1 gives the names and bounties of the target species, including harriers and buzzards, although the red kite was not grouped with these fellow raptors but instead listed alongside magpies, jays, and ravens (and priced accordingly), likely reflecting their different status and predation risk (see Chapter 5). The Tudor law that called for the mass killing of kites did give them a reprieve in urban areas, as bounties could not be collected for heads collected within two miles of a city. Under 'Exceptions' in the wording of the original act, it states that no payment would be made for the heads of any kite or raven 'killed in any City or Towne Corporate, or within two Myles of the same' (Eliz. an.8; cap.XV), perhaps indicating that the street-cleaning role kites held in medieval cities persisted to some extent in the early modern period, or at least that they were viewed as a lesser threat than they were in the countryside. The vermin laws illustrate how contextual the category of 'vermin' was, and how dependent it was on the food it eats – in contexts where animals fed on 'acceptable' food they were exempt from persecution. The urban–rural divide was cultural as well as legal, and Lovegrove (2007, 119) suggests that even if kites were beloved by medieval city-dwellers for keeping their cities clean, they were never popular with rural henwives, rabbit-shooters or farmers, who believed kites predated on their young stock. Also exempt from persecution under the vermin acts were animals that were themselves sources of food, especially for the nobility: falconry birds and their prey of herons and swans, which were prized for their prestige and entertainment value beyond simple alimentary value (see Chapter 2), along with dovecotes and rabbit warrens.

The most thorough and thoughtful scholarship on the vermin acts is provided by Lovegrove, who discusses their impacts on British wildlife (2007) including the red kite specifically (1990). Lovegrove combed parish records to understand the consequences of the vermin acts: how people perceived the animals that were now their adversaries, and the estimated ecological impacts on the populations of target species. Lovegrove's work indicates that persecution of red kites specifically was limited in the centuries immediately following the act, with a significant degree of intensification in the late-seventeenth and eighteenth centuries. Across twenty-eight English counties, no vermin payments for kites are claimed between the sixteenth and nineteenth centuries, although there are a few areas, mainly in the West Country, where kites were persecuted in higher numbers (Lovegrove 2007, 119). This suggests that although recognised as a pest and competitor for food, they were not a priority

Table 4.1 Animals listed in the Tudor vermin law of 1566 (Eliz an.8; cap.XV), in order of appearance and with stated bounty.

Animal name as given in Act	Modern equivalent (with reference if significantly different or uncertain)	Bounty (heads)	Bounty (eggs)
crowes	crow	one penny per three old or six young	one penny per six
chawghes/choughs	jackdaw (Anderson 2005; Lovegrove 2007)	one penny per three old or six young	one penny per six
pyes	magpie (1st reference)	one penny per three old or six young	one penny per six
rookes	rook	one penny per three old or six young	one penny per six
stares	starlings (Lovegrove 2007; Beazley 1914)	one penny per twelve	
martyn hawke	unknown raptor	twopence each	one penny per two
fursekytte	hen harrier (Lilford 1895); kestrel (Beazley 1914); stoat (Lovegrove 2007)	twopence each	one penny per two
moldkytte	likely unknown raptor; weasel (Lovegrove 2007)	twopence each	one penny per two
busarde	buzzard	twopence each	one penny per two
schagge	shag	twopence each	one penny per two
carmerante	cormorant	twopence each	one penny per two
ryngtayle	hen harrier (female) (Lilford 1895; Beazley 1914); harrier (Lovegrove 2007)	twopence each	one penny per two
iron	white-tailed eagle (Lovegrove 2007); heron (Beazley 1914)	fourpence each	
ospreye	osprey	fourpence each	
woodwall	woodpecker (Lovegrove 2007; Beazley 1914)	one penny each	
pye	magpie (2nd reference)	one penny each	
jaye	jay	one penny each	
raven	raven	one penny each	
kyte	kite	one penny each	
kyngfyssher	kingfisher	one penny each	

Table 4.1 (continued)

Animal name as given in Act	Modern equivalent (with reference if significantly different or uncertain)	Bounty (heads)	Bounty (eggs)
bullfynche	bullfinch	one penny each	
foxe	fox	twelvepence each	
gray	badger (Beazley 1914; Anderson 2005; Lovegrove 2007)	twelvepence each	
fitchewe	polecat (Beazley 1914); foulmart (Anderson 2005; Lovegrove 2007)	one penny each	
polcatte	polecat	one penny each	
wesell	weasel	one penny each	
stote	stoat	one penny each	
fayre bade	wildcat (Lovegrove 2007); marten (Anderson 2005)	one penny each	
wilde catte	wildcat	one penny each	
otter	otter	twopence each	
hedgehogge	hedgehog	twopence each	
rattes	rat	one penny per three	
myse	mouse	one penny per twelve	
moldwarpe	mole (Anderson 2005; Lovegrove 2007)	one halfpenny each	
wante	mole (Anderson 2005; Lovegrove 2007)	one halfpenny each	

target compared to other, more problematic vermin. As they would later be targeted in very large numbers, their absence in previous centuries indicated that there were no difficulties in sourcing and catching kites and instead their shift in representation of parish records reflects a change in pest-catching emphasis.

The legal categorisation of kites as vermin may be partially responsible for the very bad press they receive in contemporary textual sources. Like the bestiaries of early centuries, many emphasise the kite's robbing and thieving behaviour, especially in the context of food. This enmity with 'vermin' animals was also part of a wider sixteenth- and seventeenth-century theme in which nature became viewed as a sphere over which man needed to assert his dominance (Wolloch 2011). This theme was further enacted in the Enlightenment interest in the description and scientific classification of birds and animals, arguably itself a form of 'taming' (Fissell 1999). Many early natural historians did view their roles as 'cataloguing' the world around them for its potential as a resource for exploitation by humans (Raye 2021), but the category of 'vermin' became ever more clearly defined in relation to its risk to human food (Secmezsoy-Urquhart 2017). For example, the cormorant is guilty of taking not just fish but those of '3 or 4 years growth', the size preferable to humans (R.W.'s *A Necessary Family-Book both for the City & Country 1688*, quoted in Fissell 1999). In sixteenth-century naturalist William Turner's work on birds, he censured the kite's habit of taking food from humans 'unshamefastly': 'he will without any asking or begging, take away tripes and puddings from wives, whilst they are in washing of them' (Turner 1568, 54). Echoing themes found in earlier bestiaries, the kite's thieving is used as a cautionary tale against that 'which is gotten with raving and robbery, is as soon spent and wasted' (Turner 1568, 54).

Along with scavenging foods from within the sphere of human consumption, the preference for easy prey in the form of young domestic fowl made the red kite very unpopular. Francis Willughby's ornithology specifically condemns the theft of chicks from human spaces, especially hunting in 'Cities and places frequented by men; so that the very Gardens, and Courts, or Yards of houses are not secure from their ravine' (in Ray 1676, 75). For these reasons the kite is described as a particular enemy to the domestic sphere: 'our good Housewives are very angry with them, and of all birds hate and curse them most' (Ray 1676, 75). French naturalist the Comte de Buffon claims to have witnessed similar robberies first-hand, and described kites as the 'greatest tormentor and aversion' to housewives, and the 'best known' of all the 'obscene birds' (Comte de Buffon 1792). He recounts having personally witnessed kites preying on small chicks after stalking them from above, undeterred by the mother hen or by stones thrown by boys (Comte de Buffon 1792). A further domestic invasion came from the kite's habit of stealing laundry, described by Turner but also recognisable to Shakespeare's audience 'when the Kite builds, look to lesser Linen' (*The Winter's Tale*, Act 4, Scene 3; Cummings 2012). Although these texts seem to reflect a very damning public opinion of the red kite, they also emphasise that it was a familiar sight to householders and that its feeding habits continued to bring it into close (if unwelcome) contact with humans and their settlements for hundreds of years. Feeding habits are central to the negative press received by the red kite, and even their scavenging on 'non-human' foods was viewed less positively compared to other raptors: the kite is frequently compared to the hawk or eagle, whose hunting prowess was perceived as more noble. Turner (1568, 54), highlights the difference between the 'whining and lamentable pewing' of the kite and the 'noble' hawk, similar in appearance but different in behaviour. While the hawk 'feedeth upon his own prey, that he hath gotten himself', the kite is reliant on carrion provided by other predators – including humans (Turner 1568, 54). In a description of exotic cats by anatomist Nehemiah Grew, he compares the leopard to the housecat as they are similar in form but says the two 'differ, just as a Kite doth from an Eagle', with the kite as a more domesticated and less impressive bird (Grew 1685, 12). This theme is also seen in Shakespeare: 'More pity that the eagles should be mew'd / While kites and buzzards prey at liberty' (*Richard III*, Act 1, Scene 1).

Dominion over nature and food as a source of retribution

These negative perceptions would prove to be perilous for the English kite as, in the seventeenth to nineteenth centuries, persecution was intensified in a way that had not been seen in the initial few decades following the vermin acts. The earlier concerns over vermin as food competition were amplified under the ever-growing theme of mastery of man over nature, which dominated relations with the natural world in the seventeenth and eighteenth centuries. Rutkowski (2014) argues that changes to agricultural technology and economy abstracted the natural world from the 'civilised' domain of man and cities, instilling a growing sense of fear and alienation from the natural world, and perhaps this contributed towards a lack of tolerance towards kites and other animals in settled spaces. This ideological shift, as well as the specific mechanism by which animals were persecuted in accordance with the vermin laws, are both revealed in books that advised readers on how to tackle vermin. These books negotiate the food-based relationship between humans and red kites in yet another way – here food was not just the source of conflict between humans and animals, but also a means of controlling the animal and a source of danger to it, as bait. These texts marry scientific interest in the natural world with the elimination of threats and promotion of human control over disruptive animals. For example, Francis Willughby's ornithology (Ray 1676, 74–5) gives an extensive description of the kite's physical appearance, detailing the coloration of its plumage and its 'pale, but lovely yellow' eyes over several paragraphs; but in the same text also describes how to poison kites and other birds with strychnine (Ray 1676, 43). The difference between strategies for taking birds 'to eat' (hunting edible species of bird, a source of food themselves) and 'vermin' that interfered with food is clear, with dangerous or contaminating substances being used to dispatch pest species only (Ray 1676, 43). Using poisoned substances to target scavengers seems a carefully chosen retribution, as the bird's scavenging niche becomes the vehicle by which it is endangered. The birds are easily killed because they are 'noisom and ravenous', especially 'sharp-set and greedy' in the morning, and more likely to 'with greediness seize any bait' in the spring when breeding (Ray 1676, 43). The feeding habits of scavengers are also turned against them in the many forms of baited traps described in publications of this era. An early example in this genre is Mascall's book (1590, 83), which gives many 'Engines and Traps' including a 'whippe spring' baited with rabbit meat to take both kites and buzzards. In *A Necessary Family-Book* direction is given on the setting of a complex baited trap, again for both kites and buzzards, the author considering them to be equal in their 'disturbance to man' (R.W. 1688, quoted in Fissell 1999, 11–13). A century later, Smith's *Universal directory for taking alive and destroying rats, and all other kinds of four-footed and winged vermin* specifies that trapping 'the large forked-Tail kite' is best achieved using a rat or some animal intestines as bait (Smith 1786, 145). The general theme in this text and others of its type was taming and outwitting the natural world, with an emphasis on trickery and cunning; as well as catching pests, the authors often give methods for preserving fruits and other domestic arts, all falling within the realm of 'mastery of nature' (Fissell 1999). These texts were a manifestation of the Enlightenment era's triumph of man over wildness – and all the better if the animal concerned could be brought down by its own vices.

    The results of this shifting dynamic towards sovereignty over nature, combined with so many effective means to execute eradication, appear to have been ruthlessly successful. Lovegrove (2007, 120–3), in his wide review of vermin payment records, identified several parishes that seem to have undergone 'protracted campaigns' against kites specifically, such as Lezant, Cornwall, where 1,245 kites were killed between 1755 and 1809, or Bunbury, Cheshire, where payments were made for 256 heads in the single year of 1720. At Tenterden, Kent, no vermin payments were given for kites until 1654, but thirty years later a hundred birds were killed in the year 1684 alone (Ticehurst 1920). A strong change in attitude seemed to have come about in the centuries following the vermin acts, as the reframing of the red kite from tolerated to commensal to offensive pest took hold in the public imagination. People had been given a growing suite of tools and financial incentives to subdue any natural elements that were perceived to encroach on their ways of living, as well as the ideological support that it was right and just for man to rule over the natural world. Strategies of wildlife persecution were rooted in knowledge of the natural world – for example Ticehurst (1920) posits that many vermin payments were made during the nesting season, and for lots of two, three, or four birds, likely representing raiding of nests as an easier means of capture than trapping adult birds – a brutally efficient way of preventing predation on human foods.

The result of this heavy burden of persecution was the extinction of the red kite in England and Scotland by the late-nineteenth century. The final nail in the coffin was a further source of competition relating to the kite's feeding niche: large swathes of the countryside were turned over for game shooting, necessitating the preservation of game birds for sport. Predation of this stock was to be avoided, and only the fox was considered a greater enemy to this cause than wild birds of prey (Carter and Powell 2019). The shooting estate and its carefully managed stock were the ultimate countryside retreat for Victorian city-dwellers, a piece of 'desirable' or civilised nature that could not be allowed to be marred by truly wild and therefore disobedient nature, a prestige environment where high-status men could enjoy a palatable form of wildlife interaction (Ritvo 1987, 53). Lovegrove considers the kite's disappearance as testament to 'the efficiency of Hanoverian and Victorian Man's determination to obliterate species that were seen as competitors with his game-shooting interests' (Lovegrove 2007, 261). As in earlier centuries, the feeding niche of the red kite that had made them so unpopular made them easy to target: they hover and glide when seeking food and were thus easy to pick off with the latest developments in long-range guns (Carter and Powell 2019, 30). Their carrion feeding also made them vulnerable to the poisoning described by earlier authors – even when they were not the intended targets, as kites will feed on the remains of other species that have fallen victim to poisoning. Deliberate poisoning can also be shockingly effective at killing large numbers of kites, as they are relatively non-territorial and will feed collectively – so a single poisoned carcass can kill many kites (Carter and Powell 2019, 30). Kites are therefore uniquely vulnerable to various technologies employed by gamekeepers to prevent stock loss, either through deliberate targeting or as collateral damage, and this is unfortunately still true today (Smart et al. 2010; Molenaar et al. 2017). As Colonel George Hanger of Norfolk put it in 1814, while advising other sporting estate-keepers to follow his tactic of targeting kites with baited pole-traps: 'there is no animal more easily caught than the large fork-tailed kite' (Lovegrove 1990, 39).

Conservation feeding and the preservation of wildness

Food had been the cause of conflict between humans and red kites, and in many cases the mechanism behind their persecution, but in the context of changing attitudes towards wildlife in the twentieth and twenty-first centuries, it became a beneficial tool for engagement and survival. This new era of conservation included a new role for animal feeding: the provision of food in a way intended to benefit the recipient animal and promote its survival can be seen in many forms (see Chapter 11). After centuries of persecution, the red kite had maintained a precarious foothold in Wales only through careful conservation, and after more than a century of absence, they were reintroduced to England and Scotland in 1989 in a joint venture by the RSPB and the Nature Conservancy Council (Cobham 2014, 83). The chosen release areas were mosaics of open farmland and woodland combined with abundant natural food, considered to be optimal habitat (Carter and Powell 2019, 54). Nestlings were collected from donor conservationists in Sweden and Spain in areas where red kites were populous, having reached an age where birds could tear off chunks of meat when fed, allowing a more hands-off approach in their rearing, which was essential to prevent imprinting or dependency (Carter and Powell 2019, 57). After release, the birds continued to be fed on or near the release pen for some time after, as parents of young fledglings would do in the wild (Carter and Powell 2019, 58), so human feeding mimicked the role of the natural food-providing parent. The food itself was sourced from local gamekeepers and farmers, who donated carcasses of grey squirrels, corvids, and rabbits obtained through their routine pest control (Carter and Powell 2019, 57). The by-products of modern day 'vermin' control were put to use feeding a species that had only recently succeeded in crossing the barrier from 'vermin' to a once-again valued and protected animal. In this process, food was used carefully as a tool to allow kites to survive, while also preventing them from becoming attached to human feeders, and giving them the skills needed to source food when in the wild (Carter and Powell 2019, 56). In the process of reintroduction, a previously persecuted species becomes a precious resource, and the 'wildness' that was distasteful and alarming centuries ago became a quality to be protected and conserved. Although feeding was a necessary part of this, it also became a potentially polluting influence on the purity of 'wildness'. The need to maintain the separation of 'natural' and 'human' spheres may have been especially crucial since the exploitation of human food resources by the red kite had proved so fatal to them in the centuries of persecution and ecological damage.

The provision (intentional or otherwise) of food to these now-successfully reintroduced birds continued to be a key factor as their numbers grew. The monitoring of animal feeding practices is a common part of ecology, and several techniques were employed to understand how released red kites were being sustained, and how they interact with the landscape, including human sources of food. Pellet analysis is one way in which foodstuffs are identified, and Davis and Davis (1981) published a detailed account from findings in Welsh kites before the reintroductions to the rest of the UK. Their findings showed a dietary emphasis on scavenged sheep remains, including dead stock but also placentae, as well as docked tails and scrotums (evidenced by plastic docking bands), and this is anecdotally supported by farmers' accounts. Other plastic litter in nests attested to further reliance on human foodstuffs, including butcher's wrappings and cellophane packaging, although the actual food products themselves are undetectable using this method. Waste around kite nests included 'bones from butchered meat at three nests, ham or bacon rind at three, a plastic bag containing giblets from a frozen chicken, and a discarded sanitary towel' (Davis and Davis 1981). In short, the authors found that kites were considerably reliant on human foodstuffs and that exploitation of human waste products, despite being of poor nutritional value, was common practice for the kites they observed. When the diets of post-reintroduction English kites were analysed in the Midlands, sheep made up a far smaller proportion, likely because less of the land there is given over to sheep grazing and carcasses are removed far quicker (Carter and Grice 2000). In this study, pellets and fieldwork observations showed that most of the diet was scavenged carrion rather than taken as live prey, and that rabbits were especially important, although earthworms made up 20 per cent of the diet as observed by fieldworkers (Carter and Grice 2000). The authors suggest that much of the scavenged food may come from roadkill and pest-control killing, concluding that 'it is somewhat ironic that a species almost wiped out in Britain as a result of human persecution is now dependent on human support, either directly or indirectly, for much of its food' (Carter and Grice 2000, 317). Although original efforts had aimed to minimise contact between humans and birds through feeding, a reliable food supply was still needed to maintain their fragile populations. Feeding stations near release sites were established under the aegis of conservationists early in the reintroduction process. These typically began as small-scale operations by farmers who aided by providing surplus meat, but many became well-developed commercial enterprises attracting tourists with photography hides. Food now enabled people to be brought close to red kites in the same way that medieval city-dwellers had been, and provided connection and intimacy, as observed by Brettel (2016) at Bwlch Nant yr Arian, Wales. The role of the feeding station is discussed in Chapter 11 as a contributor to conservation efforts and also a nexus of connection between birds, feeding station staff, and visitors. Aspects of feeding behaviour are managed to maximise potential opportunities for visitors to connect with red kites, for example regular routine feeding enables consistent and predictable viewing events.

A more controversial form of modern-day feeding is the provision of food by individuals in domestic contexts. Although original release locations were rural, red kites flocked to locations where food was plentiful, in many cases urban or suburban settlement areas. The Chilterns release site was particularly successful in terms of population growth, with a central southern English population of 1,100 birds in a 2,600 km² area by 2016, double that of five years before (Stevens et al. 2020) and serving as a donor site for subsequent releases in the UK and abroad. Contrary to the initial aims of conservationists, who took great pains to avoid association between humans and food, the birds rapidly found abundant food supplies in the form of waste as well as food that had been deliberately provided in cities. People in release areas have taken to feeding kites directly themselves, as well as local businesses such as garden centres, and pubs and cafes with outdoor areas (Carter and Powell 2019, 120). Research in urban Reading, near the Chilterns, estimated that one in twenty households feed kites, supporting large numbers of 'commuter' kites who roost and breed out of the city but travel in for the rich pickings offered in urban gardens and around bins (Orros and Fellowes 2014 and 2015). This urban feeding can be controversial in its impacts, with some concerns about inappropriate food causing health issues in the birds, or emboldening them in attacks on picnickers to snatch food. A general anxiety about the contamination of the 'wild' status of the birds by encouraging them to come into cities and eat 'our' food underpins many of these concerns. However, despite some issues surrounding human feeding it has not proved harmful to the conservation effort overall, which has established a highly successful reintroduced population. It is also testament to the adaptability of wild animals to anthropogenic landscapes – modern cities may not be as rich in freely available offal as medieval ones, but the deliberate provision of food by city dwellers has made Reading an equally appealing habitat to kites as London once was, reopening the anthropogenic niche of the food-rich city centre.

Conclusion

The history of humans and red kites has been mediated by food in a variety of ways that have changed through time, but the overlap between their respective food niches has brought them into close and sometimes conflicting proximity. There are parallels that can be drawn between them and other commensals, other persecuted species (for instance corvids in Chapter 5) and other conservation targets, but also many unique factors relating to avian biology, ecology and cultural views surrounding birds and scavengers. They are not a domestic or tame bird, nor used as hunting animals – although predatory, they also scavenge. Critical to this relationship revolving around food is the concept of wildness – sometimes desirable to humans, sometimes not. Exploiting human food can render a wild animal 'less wild', for better or worse, or alternatively it is a means by which animals bring their wildness into urban spaces and render them uncivilised. The careful provision of food to avoid interaction with humans during the reintroduction of the red kite illustrates this, and their subsequent foraging in parks and gardens may be the first step in a new cycle from protected to pest, as they were in early modern England. The result is a careful interplay of centuries of human–wildlife interactions between urban and rural spaces, perceived spheres of 'wild' and domestic, with 'human food resources' the node around which interactions happen, neatly encapsulated in the unique case study of the British red kite. Their treatment is just one instance in which the human desire for control over nature is visible, as they have been alternately persecuted as a competing influence, encouraged through reintroduction and legal protection, or simply tolerated as a neutral commensal, but red kites are far from alone as a barometer species through which the changing attitudes of people and societies to the environment can be seen – many other examples are found in this volume such as the feral pigeon (Chapter 6) and the red fox (Chapter 10). The future of the red kite in Britain is likely to continue to be influenced by this shared food relationship, as their interest in human foods has already evoked comparisons with seagulls (for example in an article by Communications, Chilterns Conservation Board 2022), the implication being that they may earn the undesirable status of 'pest' once again. Narratives surrounding the preservation of wildness and a growing concern with how feeding birds in gardens and other spaces (see Chapter 6) impacts their populations may also affect how this feeding relationship comes to be perceived. In the case of London pigeons (Chapter 6), feeding is discouraged without reference to the preservation of wildness, implying that given large enough population numbers, a 'pest' species becomes undeserving of human food granted to scarcer, 'wilder' animals, and a narrative shift towards placing kites within this category is already visible in directives that advise against feeding them. The fluidity of animal categories and the permissibility of different types of animal feeding (utilitarian and affective) reflect changing cultural attitudes surrounding human–animal interactions, and the shifting expectations each party has regarding their respective responsibilities in maintaining a feeding relationship. For red kites in the UK, a paternalistic provision of food may contribute to a growing population, but may also drag their relationship with humans in unpredictable and dangerous directions.

References

Albarella, Umberto. 2019. *A Review of Animal Bone Evidence from Central England*. Portsmouth: Historic England.

Anderson, Douglas. 2005. '"Noyfull fowles and vermin": Parish payments for killing wildlife in Hampshire, 1533–1863', *Proceedings of the Hampshire Field Club and Archaeological Society* 60: 209–28.

Beazley, F. C. 1914. 'Notes on Shotwick', *Transactions of the Historic Society of Lancashire and Cheshire* 66 (New Series 30): 1–122.

Belon, Pierre. 1555. *L'histoire de la nature des oyseaux: Avec leurs descriptions, & naïfs portraicts retirez du naturel.* Paris: Giles Corrozet.

Brettell, Jonathan. 2016. 'Exploring the multinatural: Mobilising affect at the red kite feeding grounds, Bwlch Nant yr Arian', *Cultural Geographies* 23: 281–300.

Carr, David R. 2008. 'Controlling the butchers in late medieval English towns', *The Historian* 70 (3): 450–61.

Carter, Ian and Phil Grice. 2000. 'Studies of re-established red kites in England', *British Birds* 93: 304–22.

Carter, Ian and Dan Powell. 2019. *The Red Kite's Year*. London: Pelagic Publishing.

Ciecieznski, N. J. 2013. 'The stench of disease: Public health and the environment in late-medieval English towns and cities', *Health, Culture and Society* 4 (1): 92–104.

Clark, Willene B. 1992. *The Medieval Book of Birds: Hugh of Fouilloy's "Aviarium"*. Tempe: Arizona Center for Medieval and Renaissance Studies.

Cobham, D. 2014. *A Sparrowhawk's Lament: How British breeding birds of prey are faring*. Princeton: Princeton University Press.

Communications, Chilterns Conservation Board. 2022. 'Six reasons not to feed red kites'. Accessed 24 November 2023. https://chilterns.hapticserver.com/news/six-reasons-not-to-feed-red-kites/.

Comte de Buffon. 1792. *Buffon's Natural History, Abridged*. London: C. & G. Kearsley.

Cummings, M. J. 2012. 'Shakespeare's swooping imagery'. Accessed 2 October 2021. http://shakespearestudyguide.com/Shake2/birdsofprey.html.

Davis, P. E. and J. E. Davis. 1981. 'The food of the red kite in Wales', *Bird Study* 28: 33–40.

Fissell, Mary. 1999. 'Imagining vermin in early modern England', *History Workshop Journal* 47: 1–29.

Grew, Nehemiah. 1685. *Musaeum Regalis Societatis: Or, a catalogue and description of the natural and artificial rarities belonging to the Royal Society, and preserved at Gresham College*. London: Thomas Malthus.

Gurney, John Henry. 1920. *Early Annals of Ornithology*. London: H. F. & G. Witherby.

Holmes, Matilda. 2017. *Southern England: A review of animal remains from Saxon, medieval and post-medieval archaeological sites*. Portsmouth: Historic England.

Jones, E. L. 1972. 'The bird pests of British agriculture in recent centuries', *Agricultural History Review* 20: 107–25.

Jørgensen, Dolly. 2010a. 'What to do with waste? The challenges of waste disposal in two late medieval towns'. In *Living Cities: An anthology in urban environmental history*, edited by Matthias Legnér and Sven Lilja, 34–55. Stockholm: Forskningsrådet Formas.

Jørgensen, Dolly. 2010b. '"All good rule of the citee": Sanitation and civic government in England, 1400–1600', *Journal of Urban History* 36 (3): 300–15.

Jørgensen, Dolly. 2013. 'The medieval sense of smell, stench and sanitation'. In *Les cinq sens de la ville du Moyen Âge à nos jours*, edited by Ulrike Krampl, Robert Beck and Emmanuelle Retailllaud-Bajac, 301–13. Tours: Presses Universitaires François-Rabelais.

Jørgensen, Dolly. 2014. 'Modernity and medieval muck', *Nature and Culture* 9 (3): 225–37.

Jørgensen, Dolly. 2020. 'Crafts and cleanliness: the regulation of noxious business activity in English towns during the fourteenth through sixteenth centuries'. In *In Pursuit of Healthy Environments*, edited by Esa Ruuskanen and Heini Hakosalo, 13–26. Abingdon: Routledge.

Jørgensen, Dolly. 2021. 'Environment: managing urban sanitation for *sanitas*'. In *A Cultural History of Medicine in the Middle Ages*, Volume 2, edited by Iona McCleery, 21–38. London: Bloomsbury.

Letts, Malcolm (trans.). 1957. *The Travels of Leo of Rozmital through Germany, Flanders, England, France, Spain, Portugal and Italy 1465–1467*. London: Hakluyt Society.

Lilford, T. L. P. 1895. *Notes on the Birds of Northamptonshire and Neighbourhood*. London: R. H. Porter.

Lovegrove, Roger. 1990. *The Kite's Tale: The story of the red kite in Wales*. Sandy, Bedfordshire: Royal Society for the Protection of Birds.

Lovegrove, Roger. 2007. *Silent Fields: The long decline of a nation's wildlife*. Oxford: Oxford University Press.

Mascall, Leonard. 1590. *A booke of fishing with hooke & line, and of all other instruments thereunto belonging. Another of sundrie engines and trappes to take polcats, buzards, rattes, mice and all other kindes of vermine & beasts whatsoeuer, most profitable for all warriners, and such a delight in this kinde of sport and pastime*. London: John Wolfe.

Molenaar, Fieke M., Jenny E. Jaffe, Ian Carter, Elizabeth A. Barnett, Richard F. Shore, J. Marcus Rowcliffe and Anthony W. Sainsbury. 2017. 'Poisoning of reintroduced red kites (*Milvus milvus*) in England', *European Journal of Wildlife Research* 63: 1–8.

Mulkeen, S. and T. P. O'Connor. 1997. 'Raptors in towns: Towards an ecological model', *International Journal of Osteoarchaeology* 7: 440–9.

O'Connor, Terry 2000. 'Human refuse as a major ecological factor in medieval urban vertebrate communities'. In *Human Ecodynamics*, edited by G. Bailey, Ruth Charles and Nick Winder, 15–20. Oxford: Oxbow Books.

O'Connor, Terry. 2013. *Animals as Neighbors: The past and present of commensal animals*. East Lansing: Michigan State University Press.

Orros, Melanie E. and Mark D. E. Fellowes. 2014. 'Supplementary feeding of the reintroduced red kite *Milvus milvus* in UK gardens', *Bird Study* 61: 260–3.

Orros, Melanie E. and Mark D. E. Fellowes. 2015. 'Widespread supplementary feeding in domestic gardens explains the return of reintroduced red kites *Milvus milvus* to an urban area', *Ibis* 157 (2): 230–8.

Ray, John. 1676. *The Ornithology of Francis Willughby of Middleton in the County of Warwick Esq, Fellow of the Royal Society*. London: A. C. for John Martyn.

Raye, Lee. 2021. 'Early modern attitudes to the ravens and red kites of London', *London Journal* 46 (3): 268–83.

Ritvo, Harriet. 1987. *The Animal Estate: The English and other creatures in Victorian England*. Cambridge, MA: Harvard University Press.

Rutkowski, Paweł. 2014. '"Hurtful, venomous, ravening …": Animals as a threat in 16th- and 17th-century England. Selected examples', *Odrodzenie i Reformacja w Polsce* special issue: 181–201. https://bibliotekanauki.pl/articles/602919.pdf.

Sabine, Ernest L. 1933. 'Butchering in mediaeval London', *Speculum* 8 (3): 335–53.

Sabine, Ernest L. 1937. 'City cleaning in mediaeval London', *Speculum* 12 (1): 19–43.

Secmezsoy-Urquhart, Jessica. 2017. 'The troublesome enemy: Vermin agency in pre-modern Europe 1000–1800', *Sloth* 3 (2). https://www.animalsandsociety.org/research/sloth/sloth-volume-3-no-2-summer-2017/troublesome-enemy-vermin-agency-pre-modern-europe-1000-1800/.

Serjeantson, D. 2023. *The Archaeology of Wild Birds in Britain and Ireland*. Oxford: Oxbow Books.

Smart, Jennifer, Arjun Amar, Innes M. W. Sim, Brian Etheridge, Duncan Cameron, George Christie and Jeremy D. Wilson. 2010. 'Illegal killing slows population recovery of a re-introduced raptor of high conservation concern: The red kite *Milvus milvus*', *Biological Conservation* 143: 1278–86.

Smith, Robert. 1786. *The universal directory for taking alive and destroying rats, and all other kinds of four-footed and winged vermin*, 3rd edition. London: J. Walker.

Sneyd, Charlotte Augusta. 1847. *A Relation, or rather a True Account, of the Island of England: With sundry particulars of the customs of these people, and of the royal revenues under King Henry the Seventh. About the year 1500*. London: Camden Society.

Stevens, Matthew, Campbell Murn and Richard Hennessey. 2020. 'Population change of red kites *Milvus milvus* in central southern England between 2011 and 2016 derived from line transect surveys and multiple covariate distance sampling', *Acta Ornithologica* 54 (2): 243–54.

Taylor, John. 1630. *All the workes of Iohn Taylor the water-poet Beeing sixty and three in number. Collected into one volume by the author: with sundry new additions corrected, revised, and newly imprinted, 1630*. London: James Boler.

Ticehurst, N. F. 1920. 'On the former abundance of kite, buzzard, and raven in Kent', *British Birds* 14: 34–7.

Turner, William. 1568. *A new booke of spirituall Physik for dyverse diseases of the nobilitie and gentlemen of Englande, made by William Turner doctor of Physik*. Emden: Egidius van der Erve.

Wolloch, Nathaniel. 2011. *History and Nature in the Enlightenment: Praise of the mastery of nature in eighteenth-century historical literature*. Farnham: Ashgate.

Yalden, Derek and Umberto Albarella. 2009. *The History of British Birds*. Oxford: Oxford University Press.

5
Feed the birds but stone the crows: the role of food in conflict with corvids throughout British history
Riley Smallman

Introduction

Corvids are among the most prolific avian families around the world. Their immense adaptability and intelligence have enabled them to exploit human-modified environments (Greggor et al. 2016; Goumas et al. 2020), with a closeness between our species extending back to before the dawns of society (Ratcliffe 1997; for Neanderthal engagement with corvids see Finlayson et al. 2012). Archaeological evidence shows that corvids have been living commensally with humans for at least twenty-five thousand years (O'Connor 2013, 43; Bocheński et al. 2009). In Britain, the corvid family is represented by nine species (see Hume et al. 2020, 478–87): the northern raven (*Corvus corax*), carrion crow (*Corvus corone*), hooded crow (*Corvus cornix*), rook (*Corvus frugilegus*), western jackdaw (*Coloeus monedula*), Eurasian jay (*Garrulus glandarius*), Eurasian magpie (*Pica pica*), red-billed chough *(Pyrrhocorax pyrrhocorax),* and the rare vagrant spotted nutcracker (*Nucifraga caryocatactes*). These birds are known for their incredible intelligence (e.g. Jønsson et al. 2012; Ashton et al. 2018; Bungyar and Heinrich 2006) – with problem-solving abilities rivalling the skills of primates (Kabadayi and Osvath 2017; Sulikowski 2019; Seed et al. 2009) – as well as their long, sociable, playful lives (Marzluff and Angell 2005a, 2005b and 2012; Heinrich 1999). As generalist, omnivorous, commensal scavengers (see O'Connor 2013, 113–15; Moreno-Opo and Margalida 2013), it is food that has brought corvids close to humans – yet it is also food which has created fatal conflict between us.

Of the accusations against modern-day corvids, many centre around defending agricultural production against their scavenging, primarily crop theft and damage to livestock (Lovegrove 2007, 147–67), particularly attacks on lambs and birthing ewes (Godwin 2019). Fierce debate surrounds licences issued to shoot ravens that are perceived as threats to livestock, as ravens are a depleted and protected species in the UK (Harper 2018; BBC News 2019a, 2019b; Weston 2021; Horton 2022). Magpies also draw ire from gamekeepers, who target them to protect gamebird nests and eggs (Gough 2022; Betteley 2022). Persecutors also claim corvids' taste for eggs significantly threatens the survival of other wild birds (Swan 2022).

Despisal of corvids goes beyond the fields and into gardens and cities. Magpies and crows were found to be the second and third most strongly disliked garden birds respectively, only surpassed by disliking of wood pigeons (Cox and Gaston 2015). The study found bird species were more likeable if perceived as 'unobtrusive, brightly coloured and ... rarely a source of human–avian conflict' as well as providing interesting behaviour (Cox and Gaston 2015, 7), while species which were considered to 'out compete' others at the bird feeder garnered more contempt (Cox and Gaston 2015, 9). Corvids were also considered to be nuisance or 'disservice' species in urban environments due to perceived aggression, noise, destruction and mess (including faecal and from foraging; Cox et al. 2017, 2311): magpies due to their noise and reputation for eating songbird eggs (Cox and Gaston 2015); crows due to nesting on powerlines and 'negative perceptions by people' (Cox et al. 2017, supplementary Table S3); and jackdaws similarly for nesting locations, particularly due to causing property damage, chimney fires, and forcing people to install chimney cowls (BBC News 2003). It is notable that these three corvid species were previously described by the UK Department for Environment, Food and Rural Affairs (DEFRA) as 'pest' species, which could be killed 'for health and safety purposes', despite only one of the accusations above linking corvids with a risk to human health or safety (jackdaw nests causing chimney fires). Crows and magpies have been removed from the most recent statutory guidance on this topic, GL41 (DEFRA 2022a), however their previous inclusion indicates the level of prejudice in assuming them to be threats to human health and safety despite no clear evidence demonstrating they were. Meanwhile carrion crows, rooks, jackdaws and magpies can all be targeted to 'prevent serious damage' to livestock and crops (see GL42, DEFRA 2022b). Advice for Condition 1 under both of these statutory guidances advocate for 'reasonable endeavours' to be made towards non-lethal methods where possible – including bird-proofing and using deterrents to reduce scavenging opportunities, and utilising diversionary feeding (see Chapter 11 for definitions of feeding types and their underlying motivations) – exemplifying that food is the nexus of human–corvid conflicts.

The feeding relationship between modern humans and corvids is therefore far from utilitarian – since feeding corvids does not bring about desirable products for human usage – making it non-utilitarian in nature (for reference on these terminologies, see the Introduction). It could even be described as *anti*-utilitarian, as the unintended feeding of these birds is frequently damned as being detrimental to food supplies, livestock, gamebird and songbird populations, as well as creating interspecies urban conflict through perceived aggression, mess and noise. For many, this feeding relationship is *negatively* affective – which is to say, the accidental feeding of corvids brings some people distress and anxiety over fear of its consequences. This is in notable contrast to other forms of animal feeding discussed in this book, of which wider-reaching consequences are typically an afterthought; often only species designated as 'pests' are considered in this manner, such as Chapter 4's conflicts with red kites and Chapter 6's 'rat problems'.

This modern negativity towards corvids in Britain is notably not in line with attitudes towards them in other cultures, and – as explored in this chapter – is in stark contrast to past perceptions of corvids throughout our history. Many favourable perspectives on corvids today derive from ancient cultural and mythological roots: ravens were divine companions to the Norse god Odin (Wild 2008a, 44–6; Höfig 2007; Honegger 1998), sacred Celtic couple Sucellus and Nantosuelta (Green 1976, 11) and Greek and Roman god Apollo (Liritzis et al. 2017; Huxley 1967); corvids, including the three-legged crow Yatagarasu of Japanese myth (Knutsen 2011; Simon 2020), served as messengers and guides; deities transformed into ravens and crows, such as the Morrigan of Irish-Celtic tradition (Daimler 2020) and Japanese goddess of the sun Amaterasu (Sax 2007, 280); as well as cosmogonic creation myths, shamanic and totemic connections to corvids, such as the Rainbow Crow (Van Laan 1989) and many other indigenous legends (Von Hopffgarten 1978; Ferris 1982; Oosten and Laugrand 2006; Chowning 1962). Meanwhile, in England today, the most famous corvids are the ravens at the Tower of London – despite the historical origins of the 'tradition' of their keeping being fabricated (Sax 2007), and the contradiction between their protection as cultural icons against the backdrop of wild raven persecution. The historical and archaeological evidence unveils a pattern whereby the modern extermination and vilification of corvids is completely at odds with how they were perceived and treated throughout our sociocultural past. This change in attitudes can tell us not only about evolving relationships between people and corvids, but also about our animal feeding practices in general – or more specifically, which animals are intentionally provisioned with food, which are permitted to feed off human foods (including food waste) and which are scorned as pests to be exterminated.

New zooarchaeological research into the relationships between humans and corvids in Britain throughout history and prehistory has revealed that corvid culling is very much a 'tradition' limited to the post-medieval and modern periods. The story told within this chapter is summarised from multiple studies by the author, focusing in particular on the role of food in human–corvid interactions. These investigations have shown that corvids were once conceptualised and treated very differently from how they are today, and demonstrates how our modern negative conceptions of corvids derive from competition over and human protection of food against corvid scavenging. The carrion and waste corvids feed upon also served as a driving factor for shifting perceptions and increasing hostility towards the crow family. These new understandings of corvids in the past can help us to recalibrate our relationships with corvids in the present, focusing on the key role that food plays in this interspecies conflict.

Modern conflict and the 'tradition' of corvid persecution

Modern debates surrounding the persecution of corvids in the UK have been controversial. Farmers and gamekeepers denounce damage caused by crows, ravens and magpies in particular, stating that corvid populations need controlling for the sake of livestock and crop outputs (Swan 2022; Shute 2016; Harper 2018), while nature advocates oppose the shooting of wild birds, arguing that 'there is no evidence some of these birds, which include wood pigeons, crows, magpies and jays, are pests' (Leigh Day legal firm, in Weston 2021). Like corvids, pigeons similarly suffer from conflicting reputations and perspectives, with polarised attitudes towards urban pigeons presented within Chapter 6. In particular, *Springwatch* presenter Chris Packham has faced immense backlash for campaigning to protect corvids among other wild birds, with retaliation against him including hanging dead crows from his gate, an arson attack on his property and death threats (BBC News 2019a and 2019b; Raptor Persecution UK 2019; Godwin 2019; BBC News 2022).

However, studies into the rates of livestock lost due to corvid predation have shown attacks are rare. Houston found that only

0.12 per cent of lambs farmed in Argyll, Scotland, were damaged by hooded crows – the local abundant corvid – per year (Houston 1977, 28), and of those lambs attacked by crows, over 98 per cent of them were already extremely weak, to the point that they were expected to die regardless (Houston 1977, 27). Houston concluded 'there is no justification for control of crow numbers' on the basis of preventing ovine killings (Houston 1977, 28). Meanwhile, Bernd Heinrich, emeritus professor and raven expert, commented that 'raven bills cannot even penetrate the skin of a gray squirrel, much less the skin and skull of a sheep or a calf' (Heinrich 1999, 143). Heinrich also details a case in Germany wherein farmers were offered compensation for lambs killed by ravens, instead of shooting them: after a considerable number of claims were made, inspections were carried out, finding that most lambs had died of disease or even neglect, with ravens blamed for the killings when they had merely been scavenging the remains (Heinrich 1999, 142–5). This therefore demonstrates the desire to use ravens as scapegoats, rather than their actually being at fault: as Heinrich wrote, 'good farmers didn't have a raven problem' (Heinrich 1999, 145; for discussions on what the 'good farmer' descriptor means to people, see Chapter 7).

Other species – especially carrion crows, jays and magpies – stand accused of egg theft. While general perceptions relating to this wrongdoing stem from public concern over eggs stolen from songbird nests, theft of domestic poultry eggs is also a common complaint, as is theft of gamebird eggs on shooting estates (Gough 2022; Betteley 2022; Weidinger 2009). However, research has demonstrated that corvids predating on songbird eggs is less of a concern than commonly believed: Capstick's 2017 thesis found that increasing songbird nest mortality is primarily linked to availability of suitable nesting habitats, rather than increases in corvid populations. This reaffirms that songbird declines are more due to human destruction of environments ideal for nesting, primarily through expansion of monocultural agriculture and hedgerow removal (see Caton 2023), rather than increased predation. As with sheep slaughter, the limited impact of corvid egg theft on songbird populations is insufficient to warrant their widespread persecution (Madden et al. 2015). It may also be true that corvids are more likely to resort to egg theft when other, conflict-free food sources are less available – so driving magpies away from the birdfeeder 'in defence of songbird eggs' would more likely result in hungry corvids targeting nests.

Corvid scavenging from bins, rubbish tips and roadkill also creates repulsion towards this family of birds. Corvids benefit from their omnivorous diet and willingness to try new food sources (Greggor et al. 2016): an ample supply of human garbage is therefore a boon to the crow family, resulting in larger population accumulations wherever trash is present (Benmazouz et al. 2021; Marzluff and Angell 2005b, 292). Further dietary studies are needed to examine the health impacts of human rubbish on scavenging feeders (Plaza and Lambertucci 2017), however the impact of such feeding on human perspectives of scavenging animals is clear in our repugnance (Holmberg 2021; De Bondt and Jaffe 2022). We see our disgust towards corvid waste-scavenging exemplified in the idiom 'eating crow', which refers to the humiliation of swallowing one's pride as analogous to how deeply unpleasant it must be to eat a habitual garbage-eater.

We therefore see that many of the negative attitudes towards corvids centre around human protection of crops and livestock from scavengers, exaggeration of corvids' appetites for eggs, and revulsion towards their diets of carrion and waste; corvid diet is inextricably connected to the ways in which they are perceived and, ultimately, persecuted. It is often claimed that killing crows is a 'tradition' in the UK (Swan 2022; Lovegrove 2007, 157) – however, the antiquity of such persecution has not previously been thoroughly investigated. Zooarchaeological studies explore how and why human–animal relationships have shifted throughout history, many of which hinge on our – both human and animal – relationships with food.

Ancient corvids, gods and grains

Roman perspectives and treatments of corvids are in total disjunction with the 'persecution as tradition' narrative. Serjeantson and Morris (2011) were the first to suggest the possible ritual significance of archaeological deposits of articulated corvid skeletons from Iron Age and Roman sites in Britain, by considering contemporary mythical and cultural associations (on interpreting symbolism from animal bone depositions, see Smallman under review; Morris 2011). Ravens were the companions and messenger-birds of Apollo, whose godly realms included prophecy and agriculture – both of which were strongly connected to corvids in Roman culture (for instance, stories in Ovid's *Metamorphoses*, translated in Golding 2002, 80–4; see also Liritzis et al. 2017; Huxley 1967). Birds' calls and movements were observed as a means of divining the future in the highly respected imperial practice of ornithomancy, with ravens described as 'the only birds that seem to have any comprehension of the meaning of their auspices' (Pliny the Elder, translated in Bostock and Riley 1885, chapter 15). Note that corvids were able to give both positive and negative omens (see Arnott 2007, 109–13 on ravens, 113–16 on crows), as opposed to the typical modern supposition of them solely being harbingers of doom. Their associations with prophecy and message-deliverance can be connected to corvids' ability to speak (or more precisely, to mimic or imitate; see Marzluff and Angell 2012, 41–64): loquacious birds were extremely desirable pets, with records of ravens, crows, magpies and jackdaws being kept in Rome and further across the Empire (see in particular Lazenby 1949; Arnott 2007, 22 and 114). One story tells of a hooded crow bought by the Emperor Octavian in 31 BCE for 20,000 sesterces – approximately thirteen years' wages for the average Roman labourer (Duncan-Jones 1974; see also Scheidel 2010) – as it had been taught to say the phrase 'Hail Caesar, conquering commander' (Arnott 2007, 114–15). Pliny the Elder, in his *Naturalis Historia* (77–9 CE), describes teaching birds to speak in the Roman world:

> Birds are taught to talk in a retired spot, and where no other voice can be heard, so as to interfere with their lesson; a person sits by them, and continually repeats the words he wishes them to learn, while at the same time he encourages them by giving them food. (Pliny the Elder, translated in Bostock and Riley 1885, chapter 59)

This is notably similar to feeding as a means to bond with and train falconry birds in later periods, as shown in Chapter 2 as an intimate, intentional process: feeding is fundamental to establishing trust between person and bird. Meanwhile, in Rhodes, a tame crow was paraded in a ceremony led by the 'Crow Men' (*Korōnistai*), where the general public would offer both money and food to the crow (Arnott 2007, 114), further showing the appetite to feed these birds. Positive feeding relationships with Roman corvids also extended beyond pet care, as in accounts of attractive and diversionary feeding of wild jackdaws (see definitions of feeding types in Chapter 11):

> Aelian alleged that people in Thessaly, Lemnos and Illyria fed their Jackdaws at public expense because they made away with the eggs and young of locusts that would otherwise damage their crops … while people living at the head of the Adriatic tried to prevent Jackdaws from digging up their seed-corn by offering them barley, oil and honey cakes. (Arnott 2007, 104)

The description offered by Pliny under the title of 'A sedition that arose among the Roman people, in consequence of a raven speaking' perfectly illustrates the high esteem in which tame and talking corvids were held in the Roman capital: a raven that had taken to living in a shoemaker's store (creating a 'feeling of religious veneration' in so doing; Pliny the Elder, translated in Bostock and Riley 1885, chapter 60 (43)) would fly to the Forum and greet the Imperial family by name each day, until it was murdered by a rival store owner. The murderer was exiled from the city and later executed, while the raven was given a public funeral – typically reserved for the most prominent Roman citizens – 'with almost endless obsequies' (Pliny the Elder, translated in Bostock and Riley 1885, chapter 60 (43)). These birds were therefore highly valued and beloved, with food at the interface of such relationships.

Zooarchaeological representation of corvids in Roman Britain was higher than in any other period studied by the author (see Figure 5.1a; detailed exploration in Smallman in prep. a), with the possibility that many of these birds may have been pets or otherwise 'tame' wild corvids. Surveying artefacts from the Portable Antiquities Scheme revealed a similar pattern to the osteological remains, with the vast majority of artefacts interpreted as depicting corvids dating to the Roman period (see Figure 5.1b). The Portable Antiquities Scheme – run by the British Museum and Amgueddfa Cymru Museum Wales – records and identifies small archaeological artefacts recovered throughout England and Wales, ranging in date from Palaeolithic (Old Stone Age) to early modern, and giving insight into personally owned items and contemporary cultural symbolisms. The high representation of corvid iconography in the Roman period therefore reinforces suggestions of their sociocultural significance (further discussed in Smallman in prep. b). Many of these Roman artefacts depict corvids within the symbolic lexicon of Apollo, notably including an intaglio (engraved gem; see Figure 5.2) featuring a raven, palm branch, corn and cornucopia – all signifying agricultural prosperity (Downes 2015). We therefore see further positive associations between corvids and food abundance within Romano-British iconography.

To step backwards through time for a moment, it is also worth acknowledging the syncretism of Roman culture and religion, which frequently absorbed and rebranded aspects of different cultures encountered across the Roman Empire (Webster 1997). Many deities from other countries continued to be worshipped, becoming 'Romanised' over time – for example, Sulis Minerva, a combination of Celtic and Roman gods, venerated at the city of Bath in England (Edlund-Berry 2006; Bowman 1998). The pre-Roman divine couple Sucellus and

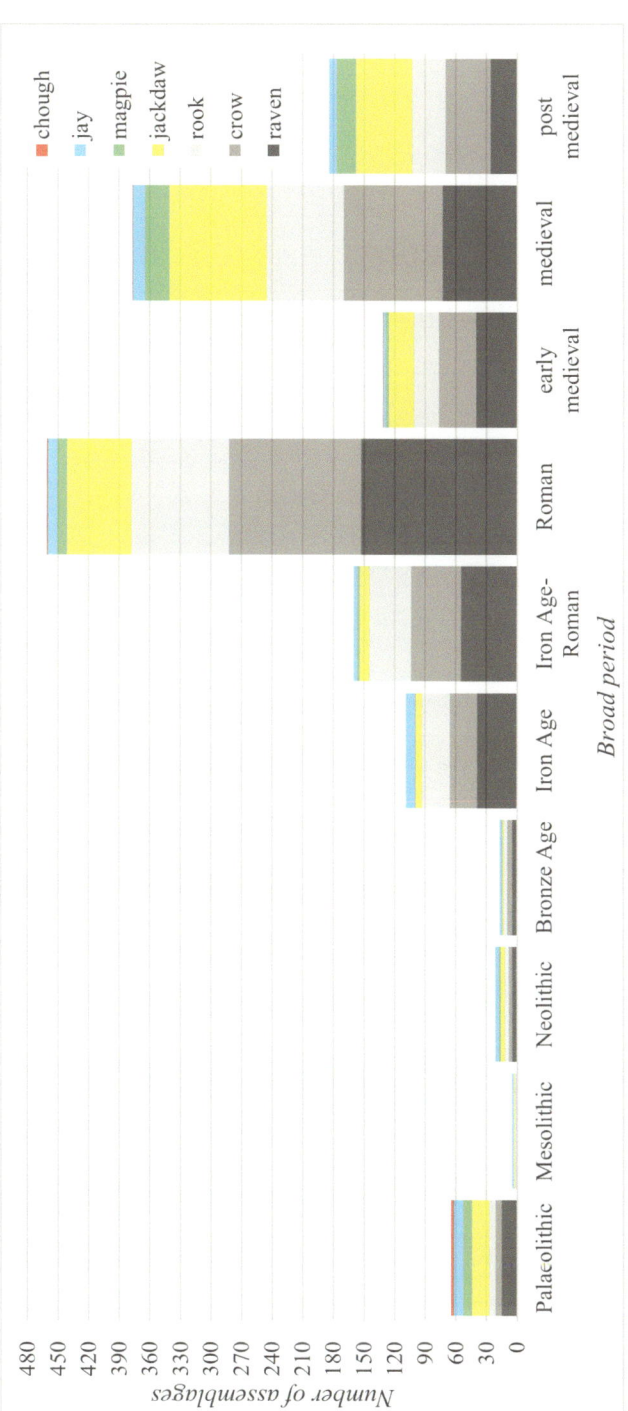

Figure 5.1a Abundance of British archaeological corvid remains, by period. *Source*: author, using data from the Archaeology Data Service and other published reviews (see Smallman in prep. a).

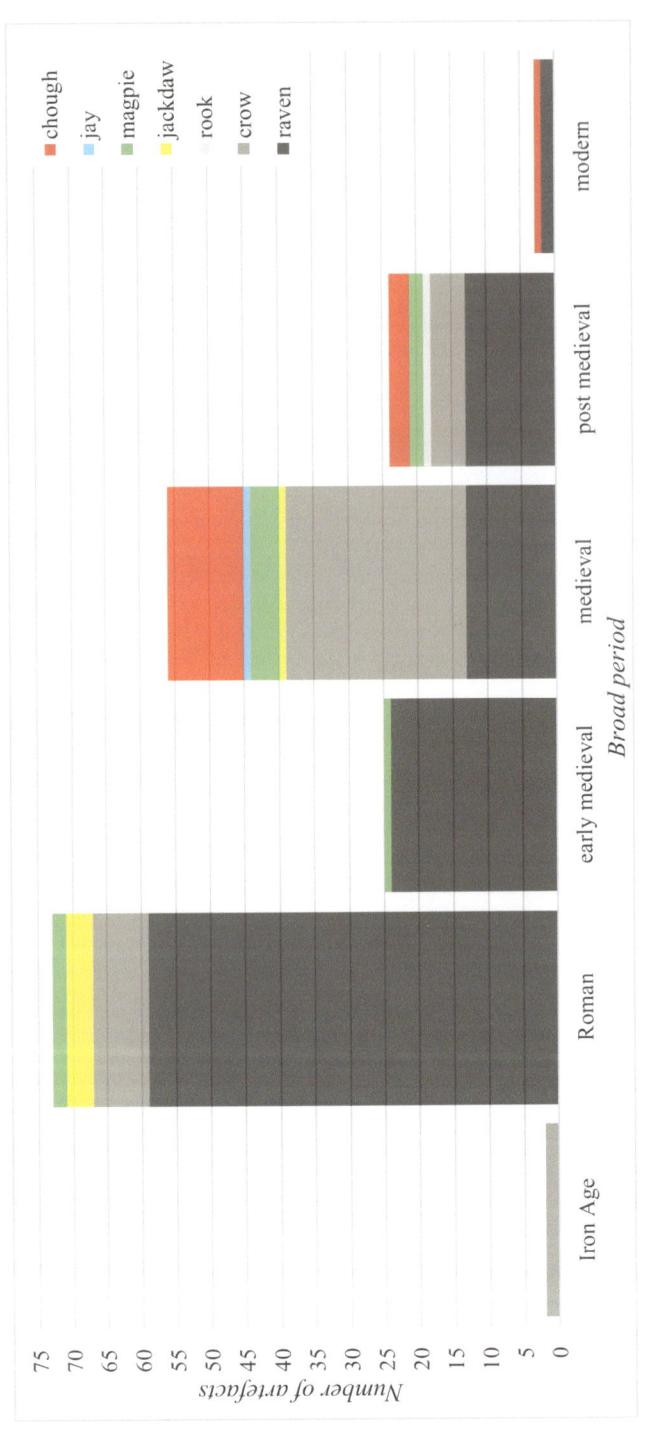

Figure 5.1b Abundance of British artefacts depicting corvids, by period. *Source*: author, using data from the Portable Antiquities Scheme.

Figure 5.2 Roman intaglio depicting raven among symbols of Apollo and agricultural prosperity. *Source*: Portable Antiquities Scheme (Record ID SWYOR-FB1233; Downes 2015; image rights West Yorkshire Archaeology Advisory Service).

Nantosuelta had ravens and crows as avian companions, with Sucellus also being a god of agriculture, much like Apollo (Green 1976, 11). Osteological representation of corvids was also high in the Iron Age and Iron Age–Roman transitionary period in Britain, although the artefactual representation was more limited (see Figure 5.1a and b), with identification of corvids in the iconography more subjective due to lack of textual evidence to verify symbolic associations. It could therefore be suggested that the perceptions and associations with ravens and crows in Roman Britain represent continuation and even reinforcement from pre-Roman ideologies – fitting with broader trends of continuity between Iron Age and Roman rituals and beliefs (Buck et al. 2019) – extending positive feeding relationships with corvids even deeper into our cultural past. This also mirrors the likelihood of deeper cultural roots of Roman veterinary and agricultural treatises, as discussed in Chapter 1.

Continuing forward, such favourable associations endured into the early medieval period, where Norse beliefs held similar mythical associations with corvids, primarily through associations with the chief god Odin (Anglicised form of the Old Norse *Óðinn*). Odin had two raven companions, Huginn (Thought) and Munnin (Memory), who would fly down to the world and report their findings back to Odin (see in particular Höfig 2007, 78–81), serving Odin in much the same way as Apollo's raven. Ravens also featured in Skaldic literature as 'beasts of battle': combining their prophetic associations and their carrion-feeding, a raven flying above soldiers heading into battle signified military success, implying the army would soon provide a feast for the ravens by slaying many enemies

(Höfig 2007, 87–9; see also Honegger 1998). This perspective is entirely antithetical to the modern repulsion towards corvids' corpse-eating habits, and allows the death association of corvids to be viewed in a totally different, valorous light. The archaeological representation of corvid bones in England declined during this period (see Figure 5.1a), however this fits with patterns across Scandinavia: ravens were not recovered from any Viking Age graves across Sweden, Norway and Denmark despite their cultural significance (Armstrong Oma under review; Karpińska 2022), potentially suggesting a cultural taboo of burying these special birds alongside humans. The portable artefact findings from England – while highly stylistic and therefore challenging to ascribe to species – generally map onto regions under Danelaw – roughly, north and east of a line from Essex to Cheshire (see Hall 1989, figure 3) – including the 'raven coin' of Óláfr Gothfrithsson, minted in York in 939–41 CE (see Wild 2008b).

Iron Age, Roman and Norse cultures in Britain therefore featured corvids in iconographic lexicons and religious beliefs, as beloved pets and deliverers of prophecy, and – in total opposition to modern perspectives – understanding them as symbols of agricultural prosperity and valorous death. However, the rooting of positive perceptions alongside religious beliefs raises the question: how did the arrival of a new dominant religion, Christianity, impact perspectives on corvids, and what role did feeding play in the shifting of attitudes towards these birds?

Crows, Christianity, crops and corpses

Christianity frequently sought to portray all symbols of 'pagan' faiths as inherently blasphemous. Christianity forbade the practice of divination, however ornithomancy was 'rebranded' as wisdom received from God via birds, such as in stories of King Solomon (Shemesh 2018a and 2018b; for other aspects rebranded into Christian narratives, see Steinforth under review). Ravens also featured in the tales of several saints: when a jealous priest tried to give Saint Benedict (in early-sixth-century Italy) a poisoned loaf of bread, Benedict asked a wild raven – whom he had befriended through feeding – to dispose of the loaf in a place where no one would find it; the raven flew away, dropping the bread into a ravine and returning three hours later (Zimmerman and Avery 1980; Standing n.d.; Barry 2019). Ravens also delivered messages and items in the seventh-century story of King/Saint Oswald of Northumbria, where a raven sent by Saint Peter helps Oswald seek a wife, among other deeds (Baker 1949; Stancliffe and Cambridge 1995; Adams 2013; Markowitz 2020;

Peterborough Cathedral 2020), as well as the ninth-century tale of Swiss hermit Saint Meinrad von Einsiedeln (Raybould 1931; Quinn 1939), where two wild ravens ensured the punishment of Meinrad's murderers. We therefore see ravens persisting as messengers in Christian narratives, including in Saint Benedict's and Meinrad's cases the intentional feeding of wild ravens to bring about a relationship of mutual trust.

However, into the second millennium CE, corvids became increasingly associated with death and corpse-eating, which were thought to be emblematic of their filth and depravity, including illustrations depicting crows and ravens feasting on human eyeballs. The *Aberdeen Bestiary* (twelfth century; see Chapter 4 for more discussion on the nature and purpose of bestiaries) includes the allegory:

> The raven picks out the eyes in corpses first, as the Devil destroys the capacity for judgement in carnal men, and proceeds to extract the brain through the eye. The raven extracts the brain through the eye, as the Devil, when it has destroyed our capacity for judgement, destroys our mental faculties. (University of Aberdeen n.d., MS 24, folio 37v)

Of the forms of corvid feeding behaviour pictured in bestiaries, one of the most common by far is based on the notion that ravens refused to feed their chicks until their plumage matured from white to black, used as a moralisation by pastors to convey the idea that they should only preach to those ready to receive their wisdom (University of Aberdeen n.d., folio 37r). The *Aberdeen Bestiary* goes on to liken the raven to a sinner who has repented and lives a life of humility (University of Aberdeen n.d., folios 37v, 38r, 38v), while the crow is praised for being an attentive parent (which goes into a lecture against abortion; University of Aberdeen n.d., folios 58r and 58v), and the jay, in contrast, is slated as a gossip, heretic or 'empty prattling' philosopher (University of Aberdeen n.d., folios 51v, 52r and 52v). Perspectives on corvids were therefore more nuanced rather than serving as a direct call to their destruction. It can therefore be suggested that Christianity largely preserved and rebranded many perspectives on corvids, however some Christian morality tales seeded notions that some species had a darker side, particularly surrounding their carrion feeding.

During the height of the Black Death (1348–9) and other epidemics and pandemics throughout the medieval and post-medieval periods, corvids scavenging human corpses was likely a common and grim sight. Plague Beaks (masks worn by doctors in outbreaks of later centuries, with a long nose holding fragrant flowers or a sponge soaked in vinegar to act as a respirator) bear striking resemblance to the black eyes and sharp bills of ravens and crows (Marzluff and Angell 2005b, 7). Corvids were also part of the morbid spectacle at execution and gibbeting sites throughout the medieval period, pecking at the hanged dead and potentially also at those left caged alive as punishment at crossroads and highways throughout the country (King 2017). Such macabre memories became inextricably linked in Britain's cultural consciousness, with the executioner's block earning the nickname of 'ravenstone' (Sax 2003) and ravens later being kept at the Tower of London as gothic props to 'dramatize accounts of executions, as proverbial birds of doom who gather at scaffolds to eat human flesh' (Sax 2007, 274). Despite corvids' increasing synonymity with corpses, accounts from travellers visiting England in the fifteenth and sixteenth centuries remarked on how tolerant the English were towards urban corvids, with some even describing fines for killing a raven (Raye 2021, 5–7). In much the same way as for the red kites discussed in Chapter 4, while these species may have been spared persecution because of their 'sanitary service' of scavenging waste – believed to prevent disease by removing decaying material that caused miasma, 'bad air' – it is unlikely they were legally protected, with clear growing distaste towards corvids' appetites expressed through Londoners' accounts (Raye 2021, 8–11). In summary, repulsion against corvids' feeding habits was softened through conceptually assigning them a utilitarian role: their usefulness made them tolerable.

The Tudor vermin laws reflect a critical juncture in the treatment and perceptions of corvids, as well as many other species, discussed more at length in Chapter 4. These laws, issued by Henry VIII and Elizabeth I in 1532 and 1566 respectively, encouraged the widespread killing of various animal species in defence of agricultural production by offering a money award for each animal killed (see Lovegrove 2007). The 1532 *Act made and ordained to destroy Choughs, Crows and Rooks* stated that 'Every One shall do his best to destroy Crows' on account of their 'marvellous destruction' of corn, grain and thatched buildings (Henry VIII an.24, cap.X, in Great Britain 1810; see also John Moore Museum 2021). A summary of the earliest recorded persecution and primary accusations against each corvid species is given in Table 5.1, the vast majority of these crimes relating to theft or destruction of human food supplies. While the earliest records of corvid persecution predate the vermin laws for many species – leading to the supposition they must have already been deeply unpopular in order to be targeted so vehemently under the laws – previous killings were on a significantly smaller scale, whereas the

Table 5.1 Summary of earliest corvid persecution and accusations by species.

Corvid Species	Earliest Recorded Persecution (all CE)	Primary Accusations
crow	~1400	sheep murder, egg theft
chough	~1400	sheep murder, grain theft, arson
rook	~1450	grain theft, noise
jackdaw	~1450	grain theft, fruit theft, inconvenient nest locations
raven	~1550	sheep murder
jay	~1600	fruit theft, egg theft
magpie	~1650	fruit theft, egg theft

Source: compiled by the author from Lovegrove 2007, 147–67.

vermin laws established attitudes and persecution efforts throughout the country.

All corvids were targeted, but none more heavily than rooks (see Lovegrove 2007, 153–8), as evidenced through the high representation of so-named 'rook and rabbit' shot on the Portable Antiquities Scheme records for the post-medieval period. In a sadly ironic turnabout, the killing of rooks in defence of grain led to the popularisation of rook pie, creating profitable demand and a trade in rook chicks. Fledglings were so valuable that poachers would steal them, leading to the saying 'shoot 'em [the poachers] hard to keep 'em [the rooks] healthy'; it was also paradoxically believed that rooks would desert a rookery if they were not regularly shot (Lovegrove 2007, 157). We therefore see those birds accused of food theft becoming popular food items themselves, with rook pie still eaten in some parts of the UK today (*Sporting Gun* 2014).

Red-billed choughs – currently the most endangered corvid in Britain, with an estimated 250–350 pairs remaining, largely restricted to cliffside Cornwall (see RSPB 2022a; Hume et al. 2020, 551), and emblematic of Cornish identity since at least the medieval period (Bullock et al. 1983; Carter et al. 2003; RSPB 2022b) – were not spared from vermin status. Just as ravens and crows are today, choughs were blamed for the 'murder' of livestock, when they were more likely just picking insects off the carcasses or from the ground around them. An additional accusation against choughs is even more tragic for its sheer infeasibility: owing to their red beaks and legs, choughs were believed to possess fire-starting abilities, and so were inexplicably damned as arsonists (Lovegrove 2007, 151). Persecution continued right up until the end of the twentieth century (Lovegrove 2007, 151–2), which, alongside the reduction of wild-grazed grasslands that they rely on for their largely insect-based diet (RSPB 2022b), led to local extinctions across England.

While the culling resulting from the Tudor vermin laws likely caused insignificant harm to national corvid populations (Lovegrove 2007, 158–9), the accusations that these laws spread dominate our perceptions and treatment of corvids to this day. Negative attitudes towards corvids likely redoubled during the British Agricultural Revolution (seventeenth to nineteenth centuries), as farmers continued to defend more efficiently produced agricultural outputs from any possible animal threat. The eighteenth century rise of gamekeeping estates undoubtedly caused severe population declines in ravens as they, among other corvids, were deemed a threat to gamebird nests for their purported penchant for eggs (Lovegrove 2007, 167). The direct impact of this persecution is clear from modern distribution and population density data, demonstrating severe reduction in raven numbers (Holyoak and Ratcliffe 1968; Gibbons et al. 1996) compared to extrapolated historical abundance (Moore 2002). Beyond the fields, the vanishing of ravens from London – the last raven pair being recorded there in 1777 – and other urban centres likely came about due to 'local factors' (Raye 2021, 11) such as increasing hygiene standards that reduced available food sources for ravens to scavenge (see also O'Connor 2013, 114), demonstrating their delicate dependency on human waste. Despite their impressive recovery (Natural England 2018), ongoing conflict still threatens the success of ravens today, particularly in eastern English counties and agricultural heartlands (see Bird Atlas Mapstore 2023; BTO 2023).

This tour through historical perceptions of corvids demonstrates clearly the relative modernness of widespread corvid persecution, as well as the centrality of food in anti-corvid sentiment. The sociocultural and climatic conditions that brought about the Tudor vermin laws bear unnerving parallels to our modern age (see Lovegrove 2007, 23–33): it is easy to understand why we are so fiercely protective of food while the world population is rapidly increasing, with developing environmental challenges and increasing risks of disease spread alongside rising political and economic pressures. However, we can challenge the historical biases informing inflammatory narratives, and try to envision mutually beneficial ways to recalibrate our modern treatment of corvids, without casting blame on those defending their livelihoods.

Ways forward: the future of human–corvid relationships

We have seen that corvids are still being targeted based on persistent accusations originating from attitudes ignited by the Tudor vermin laws. Some of these accusations are entirely fictitious (such as choughs accused of arson and killing lambs, see Lovegrove 2007, 150–3) or exaggerated and suggestive of scapegoating (corvid predation on sheep and eggs, see Houston 1977; Capstick 2017; Heinrich 1999; Shephard et al. 2015). We have also seen that people did not always hate and compete with corvids in the same way we do today: Iron Age, Roman and Norse cultures understood corvids as birds of the gods, entangled within religious and cultural identities, while early Christian narratives still often framed corvids positively. Food frequently served as a way of bringing corvids closer in the past, rather than being the wedge driven between our species. Perhaps these historical and archaeological understandings can help us to re-evaluate our relationships with corvids, particularly considering our conflicts over food.

On top of the questionable necessity of corvid culling, the effectiveness of lethal methods against corvids can also be interrogated. Removal of corvids is a great and costly effort, with further animal welfare implications (see Smith et al. 2010), while their abundance means that new birds will always replace those killed: small-scale persecution is unlikely to lead to any meaningful reduction in corvid scavenging (see Lovegrove 2007, 158–9), while total eradication would be unethical, unfeasible and unnecessary. Furthermore, corvids familiarise themselves with human faces and to some degree voices to determine potential threats, and will remember and teach other birds to recognise those who have wronged them (Cornell et al. 2012; Clucas et al. 2013; Davidson et al. 2015; McIvor et al. 2022): many anecdotes tell of corvids seeking vengeance against enemy humans and/or their property (Marzluff and Angell 2012, 169–90). Shooting corvids may then ultimately lead to escalated conflict with surviving birds, with the arms race to outsmart corvids used as a case study for cultural coevolution (Marzluff and Angell 2005a and 2005b; O'Connor 2013, 117–28) whereby both humans and non-humans continually adapt to each other's developments. Chapter 6 additionally notes how cultural coevolution is key to commensalism and feeding in shared environments. We therefore need to find a way to navigate this conflict to ensure the agricultural loss to corvids is minimised – remaining sympathetic to farmers whose livelihoods could be threatened – without further increasing rivalry and aggression between our species.

Adaptations by human cultures to reduce conflict with corvids are overviewed by Marzluff and Angell (2005b, 281–302), including caching food supplies, installing scarecrows, more secure waste bins (see also García-Arroyo et al. 2023), bird-proofing perch and nest sites with spikes and nets (but see Hiemstra et al. 2023 on corvid nests made of anti-bird spikes), percussion bird scarers in fields (see also Smith 2022) and shifting to monitored indoor lambing to protect livestock at their most vulnerable. Highlighting studies of corvid intelligence and sociability (such as Jønsson et al. 2012; Ashton et al. 2018; Bungyar and Heinrich 2006; Kabadayi and Osvath 2017; Sulikowski 2019; Marzluff and Angell 2005a, 2005b and 2012; and Heinrich 1999) also helps bring about more favourable perspectives on these birds. I propose that archaeological and historical data should be included in the toolbox for improving human–corvid interactions, by presenting how past cultures created positive relationships and symbolic attachments with corvids. Seeking understanding of how our ancestors experienced their world creates a sense of deep-time cultural identity – for example, despite the fabricated origins of the ravens at the Tower of London (see Sax 2007), these birds are a national symbol in modern England. Furthermore, we see that the horror of corpse-eating ravens has been effectively rebranded to serve as a highly successful tourist attraction (see Sax 2007), showing that it is possible to pair a feeding behaviour that would typically incite disgust with historical and cultural touchpoints to educate and entertain rather than create enmity. This non-utilitarian feeding relationship curiously combines negative and positive affective emotional responses, entertaining tourists through imaginary gore. Zooarchaeological study also critically allows us to investigate and re-evaluate how social and political biases from the past are integral to our modern viewpoints, which is particularly crucial to examine when such biases result in persecution at a level that does not fit with the facts.

It is also important to highlight the ecological role of scavenging birds (Inger et al. 2016) at a time when we are questioning human impacts on the broader environment. While corvids descending upon fields to feast on a carcass is undoubtedly disturbing – particularly to a farmer who dreads corvids killing their livestock (Lovegrove 2007, 164) – carrion-feeding species provide the vital service of removing dead material that could otherwise proliferate disease. From this perspective, utilitarian – albeit typically unwanted and uncelebrated – corvid feeding emerges in the form of corpse removal and disease prevention. However, the current system of removing dead animals sets scavenging animals as adversaries. It is the legal responsibility of UK farmers to declare dead animals and remove bodies themselves (Moore 2002, 1052): if a scavenging animal has already ripped into the carcass, this naturally makes the job more difficult and hazardous. The dynamics of such agricultural systems, and the impact on carrion-feeders as well as on our resulting perceptions of them, could therefore be drawn into further question through ecological and farming policy research.

Examining how our waste disposal systems impact other species is also vital, including the impact of rubbish dumps on species presence, behaviour and health (Plaza and Lambertucci 2017). It may even be possible to incorporate scavengers beneficially into refuse systems: in Södertälje, Sweden, a novel approach has seen crows trained to pick up cigarette butts in the street (Boffey 2022). This pilot test gave crows who did this naturally rewards of small food for each cigarette butt deposited in the receptacle. The founder of Corvid Cleaning, Günther-Hanssen, estimated trained crows could reduce the cost of cleaning cigarette butts from the street by 75 per cent (Boffey 2022), with the possibility to include other forms of litter as well. This would flip the stereotype of the 'dirty crow' on its head, encouraging the public to perceive them as helpers and cleaners as opposed to mess-makers (see Holmberg 2021; De Bondt and Jaffe 2022), reflecting ideas previously seen in the 'sanitary services' of crows in medieval England.

Meanwhile, at the bird feeder, many people are experiencing first-hand what it means to bond with wild corvids, creating positive affective relationships (BBC 2015). While typically driven away from bird feeders (Cox and Gaston 2015; Cox et al. 2017), those who welcome corvids find birds remembering them and even thanking them with small gifts (see in particular Marzluff and Angell 2012, 108–15). Some researchers have even found that wrongs can be repaired with food offerings: 'feeding or demonstrating kindness toward a crow might really be the key to winning it over' (Marzluff and Angell 2012, 190–1). This may be a sign that, while it is the protection of food that has driven us to persecute corvids, it may also be feeding corvids that brings us to new understandings and new friendships with these fascinating birds, reminiscent of relationships between people and corvids in Britain's ancient past.

References

Adams, Max. 2013. *The King in the North: The life and times of Oswald of Northumbria*. London: Head of Zeus.
Armstrong Oma, Kristin. Under review. '"Quoth the raven": on thought, memory, cognition and multispecies engagements in Scandinavian Iron Age beliefs and society'. In *Between Bones and Beliefs: Human–bird relations in central and northern Europe in the 1st millennium AD*, edited by Klaudia Karpińska, Sigmund Oehrl and Riley Smallman. Turnhout: Brepols.
Arnott, W. Geoffrey. 2007. *Birds in the Ancient World from A–Z*. Oxford: Routledge.

Ashton, Benjamin J., Amanda R. Ridley, Emily K. Edwards and Alex Thornton. 2018. 'Cognitive performance is linked to group size and affects fitness in Australian magpies', *Nature* 554: 364–7.

Baker, E. P. 1949. 'St. Oswald and his church at Zug', *Archaeologia* 93: 103–23.

Barry, Sarah. 2019. 'St. Benedict and the Raven'. Accessed 18 May 2023. https://www.stbca.org/st-benedict-and-the-raven/.

BBC News. 2003. 'Jackdaw nest kills widow'. Accessed 18 May 2023. http://news.bbc.co.uk/1/hi/wales/3226780.stm.

BBC News. 2015. 'Birds that bring gifts and do the gardening'. Accessed 13 November 2024. https://www.bbc.co.uk/news/magazine-31795681.

BBC News. 2019a. 'Chris Packham condemns dead crows hung from gate'. Accessed 18 May 2023. https://www.bbc.co.uk/news/uk-england-hampshire-48050361.

BBC News. 2019b. 'Chris Packham receives "calculated" death threat'. Accessed 18 May 2023. https://www.bbc.co.uk/news/science-environment-48105287.

BBC News. 2022. 'Chris Packham: Images released over arson attack at New Forest home'. Accessed 18 May 2023. https://www.bbc.co.uk/news/uk-england-hampshire-60758144.

Benmazouz, Isma, Jukka Jokimäki, Szabolcs Lengyel, Lajos Juhász, Marja-Liisa Kaisanlahti-Jokimäki, Gábor Kardos, Petra Paládi and László Kövér. 2021. 'Corvids in urban environments: A systematic global literature review', *Animals* 11 (3226): 1–24.

Betteley, Chris. 2022. 'Open season on shooting magpies to end'. Accessed 18 May 2023. https://www.cambrian-news.co.uk/news/open-season-on-shooting-magpies-to-end-546486.

Bird Atlas Mapstore. 2023. 'Raven'. Accessed 18 May 2023. https://app.bto.org/mapstore/StoreServlet?id=456.

Bocheński, Zbigniew M., Teresa Tomek, Jarosław Wilczyński, Jiri Svoboda, Krzysztof Wertz and Piotr Wotjal. 2009. 'Fowling during the Gravettian: The avifauna of Pavlov I, the Czech Republic', *Journal of Archaeological Science* 36: 2655–65.

Boffey, Daniel. 2022. 'Swedish firm deploys crows to pick up cigarette butts'. *The Guardian*. Accessed 18 May 2023. https://www.theguardian.com/environment/2022/feb/01/swedish-crows-pick-up-cigarette-butts-litter.

Bostock, John and H. T. Riley (trans.). 1885. *Pliny the Elder, The Natural History. Book X: The Natural History of Birds*. London: Taylor and Francis.

Bowman, Marion. 1998. 'Belief, legend and perceptions of the sacred in contemporary Bath', *Folklore* 109 (1–2): 25–31.

BTO. 2023. 'BirdFacts: Raven'. Accessed 18 May 2023. https://www.bto.org/understanding-birds/birdfacts/raven.

Buck, Trudi, Elizabeth M. Greene, Alexander Meyer, Victoria Barlow and Eleanor Graham. 2019. 'The body in the ditch: Alternative funerary practices on the northern frontier of the Roman Empire?', *Britannia* 50: 203–24.

Bullock, I. D., D. R. Drewett and S. P. Mickleburgh. 1983. 'The chough in Britain and Ireland', *British Birds* 76: 377–401.

Bungyar, Thomas and Bernd Heinrich. 2006. 'Pilfering ravens, *Corvus corax*, adjust their behaviour to social context and identity of competitors', *Animal Cognition* 9: 369–76.

Capstick, Lucy. 2017. 'Variation in the Effect of Corvid Predation on Songbird Populations', doctoral thesis, University of Exeter.

Carter, Ian, Andy Brown, Leigh Lock, Simon Wotton and Stuart Croft. 2003. 'The restoration of the red-billed chough in Cornwall', *British Birds* 96: 23–9.

Caton, Emma. 2023. 'Almost half of all UK bird species in decline'. Accessed 31 May 2024. https://www.nhm.ac.uk/discover/news/2023/april/almost-half-of-all-uk-bird-species-in-decline.html.

Chowning, Ann. 1962. 'Raven myths in northwestern North America and northeastern Asia', *Arctic Anthropology* 1 (1): 1–5.

Clucas, Barbara, John M. Marzluff, David Mackovjak and Ila Palmquist. 2013. 'Do American crows pay attention to human gaze and facial expressions?', *Ethology* 119: 296–302.

Cornell, Heather N., John M. Marzluff and Shannon Pecoraro. 2012. 'Social learning spreads knowledge about dangerous humans among American crows', *Proceedings of the Royal Society B* 279: 499–508.

Cox, Daniel T. C. and Kevin J. Gaston. 2015. 'Likeability of garden birds: Importance of species knowledge & richness in connecting people to nature', *PLOS ONE* 10 (11, e0141505): 1–14.

Cox, Daniel T.C., Hannah L. Hudson, Kate E. Plummer, Gavin M. Siriwardena, Karen Anderson, Steven Hancock, Patrick Devine-Wright and Kevin J. Gaston. 2017. 'Covariation in urban birds providing cultural services or disservices and people', *Journal of Applied Ecology* 55: 2308–19.

Daimler, Morgan. 2020. *Pagan Portals, Raven Goddess: Going deeper with the Morrigan*. Winchester: John Hunt Publishing.

Davidson, Gabrielle L., Nicola S. Clayton and Alex Thornton. 2015. 'Wild jackdaws, *Corvus monedula*, recognize individual humans and may respond to gaze direction with defensive behaviour', *Animal Behaviour* 108: 17–24.

De Bondt, Herre and Rivke Jaffe. 2022. 'Rats and sewers: Urban modernity beyond the human', *Roadsides* 8: 65–71.

DEFRA. 2022a. 'GL41: General licence to kill or take certain species of wild birds to preserve public health or public safety'. Accessed 13 November 2023. https://www.gov.uk/government/publications/wild-birds-licence-to-kill-or-take-for-public-health-or-safety-gl41/gl41-general-licence-to-kill-or-take-certain-species-of-wild-birds-to-preserve-public-health-or-public-safety.

DEFRA. 2022b. 'GL42: General licence to kill or take certain species of wild birds to prevent serious damage'. Accessed 13 November 2023. https://www.gov.uk/government/publications/wild-birds-licence-to-kill-or-take-to-prevent-serious-damage-gl42/gl42-general-licence-to-kill-or-take-certain-species-of-wild-birds-to-prevent-serious-damage.

Downes, A. 2015. 'SWYOR-FB1233: A roman intaglio'. Accessed 18 May 2023. https://finds.org.uk/database/artefacts/record/id/702135.

Duncan-Jones, Richard. 1974. *Economy of the Roman Empire*. Cambridge: Cambridge University Press.

Edlund-Berry, Ingrid. 2006. 'Hot, cold, or smelly: the power of sacred water in Roman religion, 400–100 BCE'. In *Religion in Republican Italy*, edited by Celia E. Schultz and Paul B. Harvey, 162–80. Cambridge: Cambridge University Press.

Ferris, Sean G. 1982. 'Black Power: The shamanic associations of the raven and the crow', Masters thesis, University of Windsor.

Finlayson, Clive, Kimberly Brown, Ruth Blasco, Jordi Rosell, Juan José Negro, Gary R. Bortolotti, Geraldine Finlayson, Antonio Sánchez Marco, Francisco Giles Pacheco, Joaquín Rodríguez Vidal, José S. Carrión, Darren A. Fa and José M. Rodríguez Llanes. 2012. 'Birds of a feather: Neanderthal exploitation of raptors and corvids', *PLOS One* 7 (9, e45927): 1–9.

García-Arroyo, Michelle, Miguel A. Gómez-Martínez and Ian MacGregor-Fors. 2023. 'Litter buffet: On the use of trash bins by birds in six boreal urban settlements', *Avian Research* 14 (100094): 2–9.

Gibbons, David W., Mark I. Avery and Andrew F. Brown. 1996. 'Population trends of breeding birds in the United Kingdom since 1800', *British Birds* 89: 291–305.

Godwin, Richard. 2019. 'A murder of crows: Chris Packham and the countryside war over bird killings'. *The Guardian*. Accessed 18 May 2023. https://www.theguardian.com/environment/2019/may/14/a-of-crows-chris-packham-and-the-countryside-war-over-bird-killings.

Golding, Arthur. 2002. *Ovid: Metamorphoses*. London: Penguin Books.

Gough, Andrew. 2022. 'Wild crows and magpies can now be shot so game birds can only be killed by hunters'. Accessed 18 May 2023. https://www.surgeactivism.org/articles/wild-crows-magpies-killed-general-licence.

Goumas, Madeleine, Victoria E. Lee, Neeltje J. Boogert, Laura A. Kelley and Alex Thornton. 2020. 'The role of animal cognition in human–wildlife interactions', *Frontiers in Psychology* 11 (article 589978): 1–18.

Great Britain. 1810. *The statutes of the realm: Printed by command of his majesty King George the Third, in pursuance of an address of the House of Commons of Great Britain*, Volume 3. London: Dawsons of Pall Mall.

Green, Miranda J. 1976. *A Corpus of Religious Material from the Civilian Areas of Roman Britain*. Oxford: BAR Publishing.

Greggor, Alison L., Nicola S. Clayton, Antony J. C. Fulford and Alex Thornton. 2016. 'Street smart: Faster approach towards litter in urban areas by highly neophobic corvids and less fearful birds', *Animal Behaviour* 117: 123–33.

Hall, R. A. 1989. 'The five boroughs of the Danelaw: A review of present knowledge', *Anglo-Saxon England* 18: 149–206.

Harper, M. 2018. 'A response to news that licenses have been granted to shoot ravens in England'. Accessed 18 May 2023. https://community.rspb.org.uk/ourwork/b/martinharper/posts/a-response-to-news-that-licenses-have-been-granted-to-shoot-ravens-in-england.

Heinrich, Bernd. 1999. *Mind of the Raven: Investigations and adventures with wolf-birds*. New York: HarperCollins.

Hiemstra, Auke-Florian, Cornelis W. Moeliker, Barbara Gravendeel and Menno Schilthuizen. 2023. 'Bird nests made from anti-bird spikes', *Deinsea* 21: 17–25.

Höfig, Verena Jessica. 2007. 'Raben und Rabenvögel in frühen Text- und Bildzeugnissen des Nordens'. In *Tiere in skandinavischer Literatur und Kulturgeschichte: Repräsentationsformen und Zeichenfunktionen*, edited by Annegret Heitmann, Wilhelm Heizmann and Ortrun Rehm, 73–93. Hamburg: Rombach Verlag.

Holmberg, Tora. 2021. 'Animal waste work: The case of urban sewage management in Sweden', *Contemporary Social Science* 16 (1): 14–28.

Holyoak, D. T. and D. A. Ratcliffe. 1968. 'The distribution of the raven in Britain and Ireland', *Bird Study* 15 (4): 191–7.

Honegger, Thomas. 1998. 'Form and function: The beasts of battle revisited', *English Studies* 79 (4): 289–98.

Horton, Helena. 2022: 'Ministers face legal challenge over rules for shooting wild birds'. *The Guardian*. Accessed 18 May 2023. https://www.theguardian.com/environment/2022/feb/07/ministers-face-legal-challenge-over-rules-for-shooting-wild-birds-licences.

Houston, David. 1977. 'The effect of hooded crows on hill sheep farming in Argyll, Scotland: Hooded crow damage to hill sheep', *Journal of Applied Ecology* 14 (1): 17–29.

Hume, Rob, Robert Still, Andy Swash, Hugh Harrop and David Tipling. 2020. *Britain's Birds: An identification guide to the birds of Great Britain and Ireland*. 2nd Edition. Princeton: Princeton University Press.

Huxley, George. 1967. 'White ravens', *Greek, Roman and Byzantine Studies* 8 (3): 199–202.

Inger, Richard, Daniel T.C. Cox, Esra Per, Briony A. Norton and Kevin J. Gaston. 2016. 'Ecological role of vertebrate scavengers in urban ecosystems in the UK', *Ecology and Evolution* 6: 7015–23.

John Moore Museum. 2021. '"Off with their heads!"'. Accessed 11 October 2023. https://www.johnmooremuseum.org/off-with-their-heads/.

Jønsson, Knud A., Pierre-Henri Fabre and Martin Irestedt. 2012. 'Brains, tools, innovation and biogeography in crows and ravens', *BMC Evolutionary Biology* 12 (72): 1–12.

Kabadayi, Can and Mathias Osvath. 2017. 'Ravens parallel great apes in flexible planning for tool-use and bartering', *Science* 357 (6347): 202–4.

Karpińska, Klaudia Dominika. 2022. 'On Wings to the Otherworld: Bird remains in Viking age graves from Scandinavia', doctoral thesis, University of Oslo.

King, Peter. 2017. 'Patterns of post-execution sentencing in England and Wales 1752–1834: the Murder Act in operation'. In *Punishing the Criminal Corpse, 1700–1840: Aggravated forms of the death penalty in England*, by Peter King, 77–112. London: Palgrave Macmillan.

Knutsen, Roald. 2011. *Tengu: The shamanic and esoteric origins of the Japanese martial arts*. Leiden: Brill.

Lazenby, Francis D. 1949. 'Greek and Roman household pets', *Classical Journal* 44 (5): 299–307.

Liritzis, Ioannis, Evgenia Bousoulegka, Anne Nyquist, Belen Castro, Fahad Mutlaq Alotaibi and Androniki Drivaliari. 2017. 'New evidence from archaeoastronomy on Apollo oracles and Apollo-Asclepius related cult', *Journal of Cultural Heritage* 26: 129–43.

Lovegrove, Roger. 2007. *Silent Fields: The long decline of a nation's wildlife*. Oxford: Oxford University Press.

Madden, Christine F., Beatriz Arroyo and Arjun Amar. 2015. 'A review of the impacts of corvids on bird productivity and abundance', *Ibis* 157 (1): 1–16.

Markowitz, Mike. 2020. 'Coins of the Vikings'. Accessed 18 May 2023. https://coinweek.com/ancient-coins/coinweek-ancient-coin-series-coins-of-the-vikings/.

Marzluff, John M. and Tony Angell. 2005a. 'Cultural coevolution: How the human bond with crows and ravens extends theory and raises new questions', *Journal of Ecological Anthropology* 9: 69–75.

Marzluff, John M. and Tony Angell. 2005b. *In the Company of Crows and Ravens*. New Haven, CT: Yale University Press.

Marzluff, John M. and Tony Angell. 2012. *Gifts of the Crow: How perception, emotion and thought allow smart birds to behave like humans*. New York: Atria.

McIvor, Guillam E., Victoria E. Lee and Alex Thornton. 2022. 'Nesting jackdaws' responses to human voices vary with local disturbance levels and the gender of the speaker', *Animal Behaviour* 192: 119–32.

Moore, P. G. 2002. 'Ravens (*Corvus corax corax* L.) in the British landscape: A thousand years of ecological biogeography in place-names', *Journal of Biogeography* 29: 1039–54.

Moreno-Opo, Rubén and Antoni Margalida. 2013. 'Carcasses provide resources not exclusively to scavengers: Patterns of carrion exploitation by passerine birds', *Ecosphere* 4 (8): 1–15.

Morris, J. 2011. *Investigating Animal Burials: Ritual, mundane and beyond*. Oxford: BAR Publishing.

Natural England. 2018. 'Ravens have seen impressive recovery'. Accessed 18 May 2023. https://naturalengland.blog.gov.uk/2018/07/05/ravens-have-seen-impressive-recovery/.

O'Connor, Terry. 2013. *Animals as Neighbors: The past and present of commensal species*. East Lansing: Michigan State University Press.

Oosten, Jarich and Frederic Laugrand. 2006. 'The bringer of light: The raven in Inuit tradition', *Polar Record* 42 (222): 187–204.

Peterborough Cathedral. 2020. 'King Oswald's raven'. Accessed May 18 2023. https://www.peterborough-cathedral.org.uk/oswald.aspx.

Plaza, Pablo I. and Sergio A. Lambertucci. 2017. 'How are garbage dumps impacting vertebrate demography, health, and conservation?', *Global Ecology and Conservation* 12: 9–20.

Quinn, Edward. 1939. 'Saint Meinrad', *Downside Review* 57 (1): 55–62.

Raptor Persecution UK. 2019. 'Chris Packham auctions his "penis hate mail" to support next Wild Justice legal challenge'. Accessed 18 May 2023. https://raptorpersecutionuk.org/2019/04/29/chris-packham-auctions-his-penis-hate-mail-to-support-next-wild-justice-legal-challenge/.

Ratcliffe, Derek. 1997. *The Raven*. London: T & AD Poyser.

Raybould, A. 1931. 'Einsiedeln and the shrine of the Black Madonna', *Irish Monthly* 59 (691): 22–7.

Raye, Lee. 2021. 'Early modern attitudes to the ravens and red kites of London', *London Journal* 46: 268–83.

RSPB. 2022a. 'Cornish choughs'. Accessed 18 May 2023. https://www.rspb.org.uk/birds-and-wildlife/wildlife-guides/bird-a-z/chough/cornish-choughs/.

RSPB. 2022b. 'Cornwall chough project'. Accessed 18 May 2023. https://www.rspb.org.uk/our-work/conservation/projects/cornwall-chough-project/.

Sax, Boria. 2003. *Crow*. London: Reaktion.

Sax, Boria. 2007. 'How ravens came to the Tower of London', *Society and Animals* 15: 269–83.

Scheidel, Walter. 2010. 'Roman real wages in context'. Social Science Research Network. Accessed 7 November 2024. http://dx.doi.org/10.2139/ssrn.1663559.

Seed, Amanda, Nathan Emery and Nicola Clayton. 2009. 'Intelligence in corvids and apes: A case of convergent evolution?', *Ethology* 115 (5): 401–20.

Serjeantson, Dale and James Morris. 2011. 'Ravens and crows in Iron Age and Roman Britain', *Oxford Journal of Archaeology* 30 (1): 85–107.

Shemesh, Abraham O. 2018a. '"And God gave Solomon wisdom": Proficiency in ornithomancy', *HTS Theological Studies* 74 (1): 1–9.

Shemesh, Abraham O. 2018b. '"Who tells the raven or the crane what will happen?": The Biblical prohibition of divination using birds in Classical and Medieval Jewish literature', *Journal for the Study of Religion, Nature and Culture* 12 (2): 201–24.

Shephard, T. V., S. E. G. Lea and N. Hempel de Ibarra. 2015. '"The thieving magpie"? No evidence for attraction to shiny objects', *Animal Cognition* 18: 393–7.

Shute, Joe. 2016. 'Death from above: The ravens slaughtering newborn lambs'. *The Telegraph*. Accessed 18 May 2023. https://www.telegraph.co.uk/news/2016/05/01/death-from-above-the-ravens-slaughtering-newborn-lambs/.

Simon, Scott. 2020. 'Flying the Pacific, culturing Oceania: Human–bird entanglements and Austronesian worlds', *Senri Ethnological Studies* 103: 65–87.

Smallman, Riley. Under review. 'Funerary practices most fowl: patterns of symbolic chicken deposition through Iron Age and Roman England'. In *Between Bones and Beliefs: Human bird relations in central and northern Europe in the 1st millennium AD*, edited by Klaudia Karpińska, Sigmund Oehrl and Riley Smallman. Turnhout: Brepols.

Smallman, Riley. In prep. a. 'Counting crows: The archaeological distribution of corvids in Britain from Palaeolithic to modern day'.

Smallman, Riley. In prep. b. 'Characterising corvids: Artefactual representations of corvids in Iron Age to early modern Britain'.

Smith, Jacob. 2022. 'A biosemiotic and ecoacoustic history of bird-scaring', *Biosemiotics* 15: 67–83.

Smith, Rebecca K., Andrew S. Pullin, Gavin B. Stewart and William J. Sutherland. 2010. 'Effectiveness of predator removal for enhancing bird populations', *Conservation Biology* 24 (3): 820–9.

Sporting Gun. 2014. 'Rook pie recipe'. Accessed 18 May 2023. https://www.shootinguk.co.uk/recipes/rook-pie-recipe-15406.

Stancliffe, Clare and Eric Cambridge. 1995. *Oswald: Northumbrian king to European saint*. Donington, Lincs: Paul Watkins.

Standing, Edwin Mortimer. n.d. 'Box 09, Folder 37. "The Ravens of 'Santo Speco': An ancient usage in a Benedictine monastery" (photographs included). Manuscripts, ca. 1921–ca. 1966; n.d., Edwin Mortimer Standing; 88'. Available at: https://scholarworks.seattleu.edu/standing-manuscripts/88.

Steinforth, Dirk H. Under review. 'Christian doves and Óðin's ravens: the birds of the Manx crosses'. In *Between Bones and Beliefs: Human–bird relations in central and northern Europe in the 1st millennium AD*, edited by Klaudia Karpińska, Sigmund Oehrl and Riley Smallman. Turnhout: Brepols.

Sulikowski, Danielle. 2019. 'Convergent evolution of intelligence between corvids and primates'. In *Encyclopedia of Evolutionary Psychological Science*, edited by Todd K. Shackelford and Viviana A. Weekes-Shackelford, 1434–7. New York: Springer.

Swan, Mike. 2022. 'Shooting corvids: Why our corvids are all going raven mad'. Accessed 18 May 2023. https://www.shootinguk.co.uk/shooting/shooting-corvids-why-our-corvids-are-all-going-raven-mad-132156.

University of Aberdeen. n.d. '*The Aberdeen Bestiary*: MS 24'. Accessed 18 May 2023. https://www.abdn.ac.uk/bestiary/.

Van Laan, Nancy. 1989. *Rainbow Crow*. New York: Dragonfly Books.

Von Hopffgarten, Daphne. 1978. 'The Haida Raven: A zoological and symbolic interpretation', masters thesis, University of British Columbia.

Webster, Jane. 1997. 'Necessary comparisons: A post-colonial approach to religious syncretism in the Roman provinces', *World Archaeology* 28 (3): 324–38.

Weidinger, Karel. 2009. 'Nest predators of woodland open-nesting songbirds in central Europe', *Ibis* 151 (2): 352–60.

Weston, Phoebe. 2021. 'Wild Justice: The "noisy" activists still ruffling feathers two years on'. *The Guardian*. Accessed 18 May 2023. https://www.theguardian.com/environment/2021/feb/14/wild-justice-the-noisy-activists-still-ruffling-country-feathers-two-years-on-aoe.

Wild, Leon. 2008a. 'The raven banner at Clontarf: The context of an Old Norse legendary icon'. In *Vikings and their enemies: Proceedings of a symposium held in Melbourne, 24 November 2007*, by Katrina Burge, 37–48. Melbourne: Viking Research Network.

Wild, Leon. 2008b. 'Óláfr's raven coin: Old Norse myth in circulation?', *Journal of the Australian Early Medieval Association* 4: 201–11.

Zimmermann, Odo J. and Benedict R. Avery. 1980. *Life and Miracles of St. Benedict: Book two of the Dialogues, by Pope St. Gregory the Great*. Collegeville, MN: Liturgical Press.

6
Whose food, whose health? Moral and ecological hierarchies of urban stray cats and pigeons

Giovanna Capponi and Herre de Bondt

Introduction

Animals such as cats, dogs, rats, and birds have been inhabiting urban settings since ancient times (King 2002). Despite the anthropocentric narrative of the city as a place dominated primarily by humans, recently geographers, anthropologists and human–animal scholars are increasingly reconsidering the urban environment as an 'ecological formation' (Barua and Sinha 2022) which gives place to more-than-human encounters, but also conflicts and biopolitical strategies of governance (Holmberg 2015). Urban species can occupy domestic spaces or public environments, can roam freely or be hosted in human homes, and can be protected or opposed through dedicated laws. It is important to highlight that these relational modes of existence in the urban area are not fixed. Indeed, as Chapters 4 and 5 showed us through the cases of red kites and corvids, respectively, the same animal can be loved, hated, rescued, fed or warded off by different human actors according to individual perceptions and drives towards different species. Recent anthropological work increasingly acknowledges non-human animals as actors in their own right who are not only subjected to human action but simultaneously co-create human–animal relations (Kirksey and Helmreich 2010). This framing de-emphasises the primacy of humans and instead considers the relations between humans, animals and infrastructures as constitutive of social reality as they continuously shape, reconfigure and challenge relations between each other (Barua 2021). Moreover, these relations cannot simply be analysed individually, but should be seen as part of a wider web of ecological relations that involve different species participating and inhabiting a specific place.

Inspired by relationality, in this chapter we discuss how feeding animals in urban environments creates opportunities and commensal relations not only for those species that receive direct care and attention, but also for those that inhabit the same ecological niche (Fuentes 2010). Indeed, it is shown that, in early phases of domestication, even actions performed by small groups of humans were able to foster interspecies commensalism, affecting and altering patterns of collaboration and competition between different groups. By learning to take advantage of each others' presence, animals and humans changed their preferences and behaviours, undergoing what has been defined as 'cultural coevolution' (O'Connor 2013, 118–19; Marzluff and Angell 2005). The ways in which humans cultivate and label their relations with non-humans are also subject to cultural change, so that the role and perception of certain species can change status or significance over time.

However, contemporary urban settings offer a much more complex picture: as non-human life is both regulated by governance and shaped by the action of groups who may or may not agree with the institutional understanding, the place these species should occupy in an anthropogenic setting is constantly renegotiated through moral hierarchies. Feeding often takes centre stage in urban human–animal relations; whether human provisioning of animal populations is intentional or unintentional, commensality ubiquitously nourishes relations between humans and non-human animals. Analysing this, in turn, reveals whose lives are accommodated in the city and what logics underlie this hospitality (Barua 2023). Moreover, the fostering of these commensal relations based on preferences and affinities between humans and animals creates an unexpected recombinance, meaning new ecological compositions emerge spontaneously and unexpectedly in spite of human design (Barua 2021, 1475). Commensality fosters recombinance in urban settings as feeding practices involve a variety of human and non-human actors that benefit from resources in different unexpected ways, compelling us to look at these new compositions when studying the health of humans and non-humans in cities.

We present two different case studies that exemplify these dynamics through different species and urban settings: feral cats in Rome and urban pigeons in London. In the first case, cat feeding in feral cat colonies is both spontaneously performed by cat lovers and regulated by ad hoc national laws on animal protection and animal welfare. However, the ways in which humans operate in feral cats' ecological niches create an opportunity for other species that may be subjected to different regulations or that are associated with different feelings, like gulls, crows, or insects. The second case takes us to London where, in response to, and in defiance of, anti-pigeon sentiments and actions, a group of wildlife-oriented volunteers maintain a network of vernacular care practices. Their care for pigeons includes an instrumentalised form of feeding that, as with cats in Rome, creates ecological niches and pulls in a variety of other animals. All of these animals find themselves subject to human judgements and are placed in hierarchies that humans enact through inclusive and exclusive feeding practices. The data of these different case studies have been collected using participant observation and interviews, and the authors conducted intensive fieldwork by working with interlocutors such as public health referees, veterinarians, animal welfare volunteers, cat lovers and people who enjoy feeding pigeons and birds in public spaces. Giovanna conducted fieldwork on cat colonies in Rome from February to November 2022, while Herre researched pigeon feeders in London between October 2021 and October 2023. All names are pseudonymised to maintain our respective participants' anonymity.

Feeding cats in feral cat colonies in Rome

While it is difficult to track the exact numbers, informal sources argue that Rome hosts more than four thousand feral cat colonies, of which only five hundred are legally registered, while others are informally kept and taken care of. Each of those colonies comprises between four and two hundred cats. Due to their widespread presence and importance, stray cats were declared part of the 'biocultural heritage' of the city in 2001 (Vistanet 2021). Colonies are often situated in low-traffic streets, parks, fenced archaeological sites, and monumental cemeteries, but also in marginal and neglected places. In general, feral cats define their territories in places that are not easily or continuously accessible by humans.

Stray and feral cats have always been present in the urban context of Rome. Archaeological evidence shows that, in the classical period, they were more valued as pest controllers rather than as pets, and a wide population of free-roaming and feral cats was present, especially in public spaces, outside villas, and in rural areas (Faure and Kitchener 2009, 229–30). Besides shifting to the status of pets in the Christian period, cats continued to inhabit the city of Rome and its surroundings, constituting a potential problem in terms of population control and public health.

The national law enacted in 1991 regarding feral cat colonies in Italy is described by animal welfare activists as an important achievement, also representing a unique case of urban fauna management at a national and local level. This unique legal status protects feral and stray cats from removal or relocation, highlighting the municipality's responsibility for their care, and acknowledging their legitimacy in the territory they settle into.

Despite the accusations of threatening biodiversity and spreading diseases (see the debate in Loss et al. 2012; also Lynn et al. 2019), it is important to highlight that feral cats in this context are fed primarily by humans (Natoli et al. 2022). Indeed, colonies are formed by individual cats with different stories and trajectories: some cats have spent their whole life as a stray, some have been abandoned by previous human carers, and others failed to integrate into an adoptive household. Despite displaying a variety of behaviours – from entirely feral and unapproachable to friendly and easily adoptable – cats inhabiting Roman colonies are considered to deserve care and protection just like private house cats. In fact, while in many Western and Anglo-Saxon legislations pets are protected under property laws (Favre 2017; Johnston 2021), the Italian legislation gives feral cats the right to settle in a territory and protects their lives and welfare in their own right.

Under this legal arrangement, people who feed and take care of these cats have the opportunity to voluntarily register the colony at the local public health authorities. Indeed, cat welfare volunteers, commonly known as *gattare* or 'cat ladies',[1] play a crucial role in implementing animal welfare policies. Acting as 'referees' on a voluntary basis, the cat ladies provide regular feeding and monitoring for one or more colonies, ensuring that the cats have access to food and basic care. Moreover, they help implement trap-neuter-and-return policies, for which they can claim help and reimbursement from local veterinarians or public health institutions.

The collaborative efforts between municipalities and cat welfare volunteers reflect a unique social arrangement in which the welfare of feral cats is maintained through the institutionalisation of the spontaneous action of feeding stray cats performed by cat lovers. On the one hand, this means cat lovers and volunteers have access to a series of institutional tools and guidelines while feeding stray and feral cats.

---

[1] Although some men take part in the volunteering activities, cat welfare volunteers are predominantly women, with the vernacular term *gattara* (singular) nearly always expressed as a feminine noun.

On the other hand, they develop their own empirical knowledge and personal understanding when managing different cats in different settings and ecological niches inhabited by other species. These sets of knowledge, guidelines and practices are focused on ensuring the wellbeing of a particular species – cats – whose welfare is encouraged and assisted by the law at a national and local level. However, these practices affect other species which do not maintain such a privileged status, or who are considered controversial and problematic by the institutions and the general public, creating unexpected commensal relations and affecting the health and ability to thrive of various groups of animals in the urban setting.

This short section describes how the dynamics of cat feeding in three cat colonies provide insights into both the cultural understandings of feral cats as pets and their relation to other species that gain intended or unintended access to pet food in the process.

The first colony has been registered under the name *I Gatti della Porta Magica* (The Cats of the Magic Door), located on the premises of Piazza Vittorio Emanuele II, close to the main train station of the city. Piazza Vittorio, as it is known by local residents, is a square garden that hosts, behind a protective gate, classical archaeological ruins (the Trophies of Marius) and a monument, relocated from a now-demolished seventeenth-century villa, called the 'Magic Door' due to some alchemical inscriptions. According to interviewees, a cat colony has been present in this spot since the first half of the twentieth century, and there are historical photographs of cats being fed by cat lovers since that time. Nowadays, the colony is composed of around thirty cats. Despite all being sterilised, the number of cats fluctuates due to several cases of abandonment throughout the year.

The woman in charge of taking care of this colony, Gianna, is a seventy-year-old lady who inherited her role as a carer from her mother and is now helped by a network of self-organised volunteers who take turns buying pet food and feeding the cats. The activity is regulated through a very active instant messaging group, where volunteers write daily reports on the health status of the cats, organise bulk purchases of pet food, and share pictures and news on the location. Moreover, many of the conversations in the group revolve around feeding strategies to implement in order to avoid the food being stolen by other species who inhabit the same ecological niche. Indeed, the location is also home to urban birds, such as pigeons and gulls and, during the summer, to insects such as oriental hornets. Urban birds are particularly attracted by the huge quantity of food leftovers that can be found around the square garden, where residents spend time outside, chatting, playing games and sports or picnicking. Still, due to the presence of the cat colony, pet food is also regularly available for different species to feed on.

Cat lovers ensure cats are fed twice a day: in the morning, right before sunrise, and in the evening, right after sunset. These are the times of day when the square is less busy with human residents, and also non-human ones. Indeed, the presence of diurnal birds such as pigeons and gulls make it very hard to feed cats during the daytime, as birds are quick to swoop in and eat the food.

The feeding strategies are organised based on seasonality and temporality. In fact, the main gate of the square garden closes earlier during the winter, while the location remains accessible for longer during the summer. Moreover, cats need to be fed in cooler spots of the square when the heat is more intense or in sheltered spots when it is raining. Therefore, feeding activities are planned by observing various shifting criteria: the cats' movements, the weather, daylight hours and the movements and habits of other species that may take advantage of the food.

On an average shift, cat volunteers get in front of the gated area at the set time and start calling the cats. Some cats are normally already waiting for food, while others emerge from their hiding spots when hearing the call. The cat volunteer then prepares several plastic plates with a mixture of dry and canned pet food (see Figure 6.1). Dominant or stronger cats normally eat first, while the others wait their turn. During the whole process, volunteers stay there to make sure that each cat receives a portion of food by moving the plates around closer to the shyer cats. Moreover, volunteers must keep away the gulls active in the area by waving their hands or making noises. In fact, during the whole process gulls try to get closer to the food or wait for a plate to be left unattended (either by the cats or by the volunteers) so as to steal the leftovers. During the summer, the food also attracts oriental hornets, big insects who prevent the cats from eating peacefully by flying and buzzing around the plates. When these competitor species become particularly problematic, volunteers adopt strategies such as separating a small plate of food for them to feed on so as to keep them away from the cats.

Indeed, while cat lovers are mostly preoccupied with the need for the cats to have regular access to food, they also seem to believe that other species deserve their share: at the end of the shift, when all cats have eaten, the leftovers are not thrown away, but put in a corner where birds can feed on them. This pattern shows different intentionalities

Figure 6.1 A stray cat being fed at the colony of the Magic Door, Piazza Vittorio Emanuele II, Rome. *Source*: author (Capponi).

around feeding practices. On the one hand, cats are given full priority and are fed with good quality pet food. This means they are regarded as more vulnerable than birds when it comes to foraging or hunting. Indeed, birds are believed to be able to take care of themselves as feral/wild species. On the other hand, the regularity of feeding patterns ensures that the cats stay in the area instead of looking for food somewhere else, which would imply road-crossing and other dangerous situations. Regular feeding is not only a strategy to keep cats healthy by human standards, but also to keep them physically within the territory of the colony. Ultimately, these hierarchical patterns in feeding show a mixture of care and control. In fact, providing food to urban fauna is not only motivated by feelings of affection and concern towards cats, but also by a sense of stewardship in which humans tend to direct animals' movements and presence in a desired way.

Interestingly, these patterns can be observed also in other contexts, such as enclosed cat shelters. One case is represented by the cat shelter of Azalea, an organisation located in the San Camillo Forlanini hospital, in the Gianicolense neighbourhood, southwest of the city centre. Azalea is a private association with a special partnership with the municipality of Rome. It hosts almost three hundred cats and is sustained through both public funds and donations. The location had originally been the territory of a cat colony that spontaneously settled in the garden of the hospital. In the early 2000s it was then fenced and converted into a shelter, where cats with various health conditions receive food and medical care. Cats are fed twice a day with canned pet food, while dry food is available to them throughout the day. During the feeding times, volunteers are instructed to clean the leftovers and to check if the dry pet food is stale or mouldy, replacing it as necessary. These leftovers are not thrown away but set aside for the crows and smaller birds living in the hospital's garden, right outside the shelter. In general, these practices are motivated by strong support for the idea of recycling resources and materials. Indeed, in cat shelters old blankets and jumpers are used to make the kennels more comfortable, all sorts of containers are reused to store food, and plastic bottles are cut to make scoops and other utensils. By the same token volunteers argue that food is also recycled by being given to birds, which, in their understanding, deserve food and care like any other living being. The way in which cat volunteers share these resources makes it advantageous for flocks of crows and garden birds to keep dwelling around the cat shelter.

A similar case is exemplified by the cat colony of Villa De Sanctis, a park in the district of Tor Pignattara, in the east side of the city. The colony is composed of a small group of around five cats living primarily outside. However, the cat volunteers have been given permission to build a small enclosed shelter inside the park, where they keep cats who need special treatment or suffer health issues. Luciana, the main referee of the colony, is an active member in the local community, and she also volunteers in the local food bank. Every day she comes to feed the cats in the colony and in the shelter, mixing it with the appropriate treatments and medicines. Since running water is not available in the shelter, cat volunteers bring water in tanks from the closest fountain to fill the cats' water bowls with clean drinking water. However, dirty water remains are not thrown away but poured into a bucket with leftover stale bread, which Luciana brings from the food bank. The stale bread is soaked until soft, and cat volunteers crumble it with their hands and toss it to the crows, pigeons and parakeets living in the park, just outside the shelter. Here, the recycled food that is shared with birds is composed of a variety of human and non-human leftovers: stale bread from the food bank and water from the bowls of feral cats.

In this section we have shown that feral and stray cats in Rome benefit from a special status, their rights to be protected as non-human denizens have legal basis, and 'cat ladies' collaborate with public health institutions as official gatekeepers of the cat colonies. Feeding practices take into account the presence of other species, creating commensal relations with them, but based on clear moral hierarchies. While cats have privileged treatment both at an institutional and practical level, feeding gives place to new recombinant ecological formations, which also benefit other species in various ways. However, what happens when the target species are not what institutions expect, and undesirable species are targeted as deserving food and care?

Pigeon feeding in London

The Museum of London offers visitors a variety of London-themed gifts and souvenirs in their shop, of which a significant number depict rock pigeons (*Columba livia*), suggesting they are meaningful to London. Presenting the feral rock pigeon as an inhabitant of the city and commodifying this image by selling pigeon-themed items from Christmas decorations to jigsaw puzzles contrasts the contested position of these birds. Once domesticated for their meat, aesthetic quality and homing capacities, pigeons feralised and became a well-established urban resident in the modern city. London's built environment provides pigeons with a space in which they can thrive by roosting and nesting on architecture that resembles the cliff faces their ancestors once inhabited (Jarvis 2011). Aside from their roosting and nesting habits, pigeons' ways of foraging and feeding – being synanthropic ground feeders – renders them non-human pedestrians that share sidewalks, streets and squares with the terrestrial human. Initially they were primarily fed by spillage of the grain that was used to feed horses drawing carriages

(Barua 2021), but pigeons now thrive wherever edible food waste is available (Buijs and Van Wijnen 2001; Jokimäki and Suhonen 1998). London has no shortage of human residents, and even though only a small portion of Londoners intentionally feed pigeons, humans are messy eaters whose continuous supply of waste becomes currency of bargain with birds (Smyth 2020), incentivising pigeons and other synanthropic species to stay in cities. Thus, history, architecture and human feeding practices inextricably connect pigeons and humans in contemporary London, where both species' lives continuously intersect.

The encounters that this intersection of species generates result in conflict, however, as the birds are thought to pose a material, cultural and symbolic threat to the city. Their faecal matter threatens material integrity of the built environment, their bodies pose a perceived health risk and, to many, the feral pigeon symbolises the filthy and contaminated dimensions of the urban environment. Through their presence in the city as well as their agency, pigeons threaten the anthropogenic design of cities and challenge the idea that nature should be subservient to humankind. They serve as a reminder that cities find themselves in a constant flux, as all manner of beings – human and non-human alike – are continually making and reconfiguring the city according to their own competencies and rhythms (Barua 2023, 4). The government-led pigeon ban in Trafalgar Square at the beginning of the twenty-first century attests to the contentiousness of this bird's presence in the city (BBC 2007). Mayor Ken Livingstone deemed pigeons unfitting of his vision of the square as a 'cultural space' and decided to ban the square's bird-feed vendors, fine any feeders and employ a company to fly Harris hawks to deter pigeons. In the meantime, opponents of Livingstone's plans doubled down on feeding the remaining population of the square to minimise their suffering related to this project (Escobar 2014). Pigeons find themselves in the middle of a human debate about ecology, health and morality, resulting in normative claims about which animals should be fed and which animals should actively be prevented from being fed.

Despite widespread objections from councils, organisations and members of the public, pigeon feeding remains a ubiquitous pastime for many Londoners. Apart from unintentional feeding through litter and the occasional scrap, a significant number of human residents go out of their way to feed pigeons in a focused manner (Jerolmack 2013). Motivations for bird feeding differ significantly: one might simply enjoy the interaction with non-human others, while others feed to maintain the health of pigeon populations. Herre conducted ethnographic fieldwork with the latter group of pigeon feeders in London as he joined a group of wildlife volunteers in feeding pigeons in an attempt to catch them and administer medical care. Members of London Wildlife Protection and the affiliated 'Stringfoot Sunday' (SFS) community commit themselves to catching, rescuing and rehabilitating the city's pigeons that are in dire need of help. String and hair wrapped around their feet (Jiguet et al. 2019), paramyxovirus, trichomoniasis and other internal or external injuries pose a salient threat to urban rock pigeons. This wildlife-minded group of Londoners aims to minimise these constant threats to pigeons, motivated by reaction against the hard-to-miss anti-pigeon sentiments and actions and by a firm belief that every life matters. Feeding, in this context, takes a more instrumental role: the primary aim is not to provide supplementary food, but to use this food to catch and provide care for injured pigeons. Nevertheless, feeding remains an essential part of maintaining a healthy population of pigeons.

The locations where pigeons and feeders interact become nutritional hotspots, and this surge in food raises the question: what is feeding doing *beyond* the pigeon? Feeding events create their own ecological niches as a variety of animals learn where and when provisioning takes place and deliberately frequent such sites before, during and after people feed pigeons. Each of these animals relate differently to the targeted pigeons, to feeding humans and to the wider urban ecology. Perhaps the most contested example of this form of commensalism is the brown rat. Various signs in London's parks reveal that feeding practices are not only feeding pigeons, but that this provision of food simultaneously feeds the public anxieties regarding rats. Research in Amsterdam concluded that bird feeding in public spaces sustains up to 153 rats per feeding location (Burt et al. 2020), particularly when people feed later in the day and food is more likely to be left by satiated birds for the mainly nocturnal rat to scavenge on later. While Amsterdam and London are not the same, the morphological, infrastructural and cultural similarities make it safe to assume that feeding is sustaining rat populations of similar size in London. Ever since New York's parks commissioner Thomas P. Hoving dubbed them 'rats with wings' (*New York Times* 1966), humans equate pigeons with rats as they are both thought to consume the refuse of society (Jerolmack 2008). Both species are generalist scavengers that thrive by foraging for urban litter while residing in close proximity to humans, and feeding effectively reinforces this similarity as both pigeons and rats feed on any leftover grain, bread or other feeding material.

Since a significant portion of bird feeding takes place in parks, park management organisations often take a clear position regarding feeding practices, which they communicate to the public through signage. Royal Parks oversees eight of London's largest open spaces, which includes the management of wildlife-related issues that emerge in the capital's parks. The parks' signage, and messages on the Royal Parks website, inform the general public that a certain hierarchy of animals is in place in the parks. While signs discouraging pigeon feeding merely implore the reader not to feed pigeons as it is prohibited, signs asking the reader not to feed ducks, geese or other waterfowl legitimise their request by arguing that it is for the sake of the animals' health, whereas pigeon health rarely features as an argument. Similarly, the Royal Parks' 'Keep wildlife wild' initiative (Royal Parks n.d.), as well as the guidelines on feeding (Royal Parks 2020) reveal normative hierarchies heralding and welcoming conventional charismatic species (Lorimer 2007) – such as geese, herons and deer – while speaking of deterring animals that are argued to pose threats to biodiversity health – such as crows, magpies, gulls, squirrels and rats. The park aims to limit feeding altogether in their efforts to 'keep wildlife wild' and to protect vegetation and water quality, but their legitimations reveal normative feeding logics that favour some animals over others.

Similar signage found in Waterlow Park, in the green and hilly area of Highgate (north London), reveals that institutions similarly acknowledge the creation of ecological niches that emerge when people feed certain animals. The Parks and Green Spaces department of Camden Council placed a sign reading 'PLEASE Do not feed the birds' followed by an illustration of a rat and the pay-off line 'Because it's not only the birds you are feeding!' (Figure 6.2). By including this last line, they convey to the reader that their feeding practices may attract rats, but the lack of elaboration reveals that they expect the reader to understand and agree that this is an undesirable outcome. In fact, sustaining a rat population in the park is so objectionable that the Green Spaces department would rather deter visitors from feeding birds entirely if that serves the goal of preventing rats. Moreover, the fact that, in colloquial language, pigeons are referred to as 'rats with wings', while rats are not referred to as 'crawling pigeons', implies that, while both species are stigmatised for being destructive, ubiquitous and hazardous to public health, pigeons occupy a higher rank in the feeding hierarchy.

One of Herre's interlocutors who lives near Waterlow Park not only feeds pigeons daily, but also openly expresses her disapproval of the Parks and Green Spaces department through vandalism. Herre's interlocutor 'enhanced' this sign by adding and removing letters with a pen, making it say 'PLEASE Do not forget to feed the birds', followed by a drawing of a heart. She then anthropomorphised the rat by drawing a smiling mouth besides which she wrote 'I also live here in the park', attributing those words to the rat. The interlocutor further added letters to the contact information, making it read 'Parks and Green Spaces – That's what they're for! Enjoy them' (Figure 6.2). She proudly shared her artistic reimagining of the sign with the Stringfoot Sunday group, correctly expecting it to be met with approval. This act of vandalism reveals how belonging is disputed as official channels actively discourage feeding due to the untargeted species that it attracts, while London's feeders embrace and celebrate the creation of ecological niches as a by-product of their pigeon-feeding habits. On multiple occasions Stringfoot Sunday members expressed their belief that all life matters and that this belief fuels their feeding practices. Additionally, this institutional discouragement and general disdain of species that most people deem undesirable is exactly what galvanizes SFS members to keep feeding, as they feel sorry for rats, pigeons and corvids.

SFS members express the view that 'all life matters', and their feeding practices reflect this to some extent as they include far more than just pigeons. Their primary reason for feeding may be to catch and care for pigeons, but their feeding creates ecological niches that attract a multitude of other animals. That all life matters becomes evident in how

Figure 6.2 Sign in Waterlow Park posted by Camden Council (left). Same sign after alterations by SFS member (right). *Source*: anonymous interlocutor, used with permission.

they cope with the influx of geese, ducks, crows and other birds. Julie, an SFS member living in Canada Water (in south-east London), brought Herre to Southwark Park, where a patch of grass bordering a pond makes up this member's feeding grounds. Herre and Julie arrived in the park with a stroller carrying not only Julie's dog Rambo but also a plastic bag of oats, a bag of frozen peas, red peppers and a large bag of mixed bird seed. While feeding pigeons is the main goal of Julie's trip to Southwark Park, she anticipated other avian guests as well, as the birds of the park have come to recognize her. Before turning to the pigeons, Julie rips open the bag of peas and empties it in the water for the Canada geese and ducks that have gathered around, she divides the bag of oats in small piles in a circular pattern that the crows feast on, and she breaks and tosses bits of pepper towards the same crows. It is only when all other birds are preoccupied that she turns to the pigeons that have stayed at a safe distance behind the iron fencing and starts to throw them the mixed seeds. This feeding strategy not only supports a greater variety of urban wildlife that her feeding habits attract, but it simultaneously allows her to care more attentively for the pigeons. In groups of other bird species, pigeons often find themselves low on the pecking order and are ousted by the more assertive geese or crows if they are all targeting the same food source at the same time.

Feeding creates ecological niches, and this comes with a level of uncertainty that institutions are uncomfortable with; untargeted, unexpected and undesirable animals may emerge and could potentially have a detrimental effect on other animals, humans or on the park itself. Feeders like Julie directly challenge the moral hierarchy that institutions communicate. This hierarchy places pigeons fairly low – albeit just above rats – and determines that pigeons should not receive food, let alone care. When SFS members practice care for feral pigeons in the form of feeding and medical aid, recombinance inevitably happens as nearby animals make use of the feeding opportunity. Feeding these non-targeted species deliberately serves as a distraction and allows for more attentive care for pigeons, and in this process SFS members regard care for non-targeted species as an inevitable consequence and responsibility of feeding. While the normative, top-down perspective tells us to limit the creation of ecological niches to prevent recombinance's unpredictability, pigeon feeders teach us to abandon the illusion of control and to embrace the unexpected instead.

Conclusion

Rome's cats and London's pigeons find themselves in a remarkably similar position: both populations were once domesticated for human purposes but have since broken out of this position. Not quite domesticated due to their spatial, reproductive and behavioural freedom, and yet not quite wild due to their reliance on anthropogenic food sources, cats and pigeons both arguably occupy the liminal position of feral animals (Holmberg 2014, 56–7). Despite both cats and pigeons following a similar pathway from domestic to feral, cats in Rome are met with state-endorsed feeding and even subsidies, while London's pigeons are met with municipally endorsed hostility and anti-feeding measures. In effect, Rome's feral cats continue to enjoy the status of pets, deserving priority in receiving food and care. By contrast pigeons have been reconfigured as pests, so that feeding and treating them has been discouraged and criminalised. Despite these differences, both case studies confirm that humans have a proclivity to feed animals whatever the financial cost, legal repercussions, time investment or other barriers. As we can see in Chapters 1 and 7, feeding is an ostensibly utilitarian practice that serves to sustain animal populations, while simultaneously fraught with – and motivated by – affective sentiments. Rome's 'cat ladies' organise themselves around this feeding activity on a voluntary basis, often investing their own resources in high-quality food and medical care for the cats, while London's pigeon feeders similarly spend significant amounts of time, money and effort on providing pigeons with supplementary food.

Techniques of trust, control and care are co-constructed with animals and their agency, taking into account animals' behaviours and preferences, but also developing a greater awareness of the wider urban niche and its inhabitants. Indeed, it has been observed how feeding practices create unexpected availability of resources for a variety of species who interact in the same space. However, resources are not accessed evenly, but administered and distributed according to social and cultural understandings of what and how animals should deserve care. In this way, analysing feeding practices and strategies not only mirror the different places animals occupy in human moral hierarchies, but also create new unintended commensal relations and recombinant interactions among species. As a result, these practices cannot be understood simply as a relation between human feeders and carers and their target species, but as actions that deeply affect the whole ecological niche. Moreover, as Rome's cats and London's pigeons show us, we should not only acknowledge these recombinant forms of urban nature as a *result* of feeding, but as a driving force in *shaping* feeding practices. Whether it is Rome's cat volunteers feeding cats at specific times to avoid certain other species or London's pigeon feeders bringing specific food items to distract other potentially disruptive birds, recombinance continuously reshapes how we feed.

References

Barua, Maan. 2021. 'Infrastructure and non-human life: A wider ontology', *Progress in Human Geography* 45: 1467–89.
Barua, Maan. 2023. *Lively Cities: Reconfiguring urban ecology*. Minneapolis: University of Minnesota Press.
Barua, Maan and Anindya Sinha. 2022. 'Cultivated, feral, wild: The urban as an ecological formation', *Urban Geography* 44 (10): 2206–27.
BBC. 2007. 'Trafalgar's pigeon ban extended'. Accessed 2 November 2023. http://news.bbc.co.uk/1/hi/england/london/6986166.stm.
Buijs, Jan A. and Joop H. Van Wijnen. 2001. 'Survey of feral rock doves (*Columba livia*) in Amsterdam, a bird–human association', *Urban Ecosystems* 5 (4): 235–41.
Burt, Sara A., Cornelis J. Vos, Jan A. Buijs and Ronald J. Corbee. 2020. 'Nutritional implications of feeding free-living birds in public urban areas', *Journal of Animal Physiology and Animal Nutrition* 105 (2): 385–93.
Escobar, Marie Paula. 2014. 'The power of (dis)placement: Pigeons and urban regeneration in Trafalgar Square', *Cultural Geographies* 21 (3): 363–87.
Faure, Eric and Andrew C. Kitchener. 2009. 'An archaeological and historical review of the relationships between felids and people', *Anthrozoös* 22: 221–38.
Favre, David. 2017. 'Animals as living property'. In *The Oxford Handbook of Animal Studies*, edited by Linda Kalof, 65–80. Oxford: Oxford University Press.
Fuentes, Agustín. 2010. 'Naturalcultural encounters in Bali: Monkeys, temples, tourists, and ethnoprimatology', *Cultural Anthropology* 25 (4): 600–24.
Holmberg, Tora. 2014. 'Wherever I lay my cat: post-human crowding and the meaning of home'. In *The Routledge Handbook of Human–Animal Studies*, edited by Garry Marvin and Susan McHugh, 54–67. London: Routledge.
Holmberg, Tora. 2015. *Urban Animals: Crowding in zoocities*. London: Routledge.
Jarvis, Peter J. 2011. 'Feral animals in the urban environment'. In *The Routledge Handbook of Urban Ecology*, edited by Ian Douglas, David Goode, Mike Houck and Rusong Wang, 361–9. Abingdon: Routledge.
Jerolmack, Colin. 2008. 'How pigeons became rats: The cultural-spatial logic of problem animals', *Social Problems* 55 (1): 72–94.
Jerolmack, Colin. 2013. *The Global Pigeon*. Chicago: University of Chicago Press.
Jiguet, Frédéric, Linda Sunnen, Anne-Caroline Prévot and Karine Princé. 2019. 'Urban pigeons losing toes due to human activities', *Biological Conservation* 240 (108241): 1–6.
Johnston, Jacquelyn. 2021. 'Incongruous killing: Cats, nonhuman resistance, and precarious life beyond biopolitical techniques of making-live', *Contemporary Social Science* 16 (1): 71–83.
Jokimäki, Jukka and Jukka Suhonen. 1998. 'Distribution and habitat selection of wintering birds in urban environments', *Landscape and Urban Planning* 34: 253–63.
King, Anthony. 2002. 'Mammals: evidence from wall paintings, sculpture, mosaics, faunal remains, and ancient literary sources'. In *The Natural History of Pompeii*, edited by Wilhelmina Feemster Jashemski and Frederick G. Meyer, 401–50. Cambridge: Cambridge University Press.
Kirksey, S. Eben and Stefan Helmreich. 2010. 'The emergence of multispecies ethnography', *Cultural Anthropology* 25 (4): 545–76.
Lorimer, Jamie. 2007. 'Nonhuman charisma', *Environment and Planning D: Society and Space*, 25 (5): 911–32.

Loss, Scott R., Tom Will and Peter P. Marra. 2012. 'The impact of free-ranging domestic cats on wildlife of the United States', *Nature Communications* 4: 1–8.

Lynn, William S., Francisco Santiago-Ávila, Joann Lindenmayer, John Hadidian, Arian Wallach and Barbara J. King. 2019. 'A moral panic over cats', *Conservation Biology* 33: 769–76.

Marzluff, John M. and Tony Angell. 2005. 'Cultural coevolution: How the human bond with crows and ravens extends theory and raises new questions', *Journal of Ecological Anthropology* 9: 69–75.

Natoli, Eugenia, Carla Litchfield and Dominique Pontier. 2022. 'Coexistence between humans and "misunderstood" domestic cats in the anthropocene: Exploring behavioural plasticity as a gatekeeper of evolution', *Animals* 12 (13): 1–20.

New York Times. 1966. 'Hoving calls a meeting to plan for restoration of Bryant Park; cleanup is urged for Bryant Park'. Accessed 23 July 2023. https://www.nytimes.com/1966/06/22/archives/hoving-calls-a-meeting-to-plan-for-restoration-of-bryant-park.html.

O'Connor, Terry. 2013. *Animals as Neighbors: The past and present of commensal animals*. East Lansing: Michigan State University Press.

Royal Parks. 2020. *The Royal Parks Biodiversity Framework (2020–2030)*. London: The Royal Parks.

Royal Parks. n.d. 'Keep wildlife wild'. Accessed 23 July 2023. https://www.royalparks.org.uk/help-nature-thrive/how-to-get-involved/keep-wildlife-wild.

Smyth, Richard. 2020. *An Indifference of Birds*. Axminster: Uniformbooks.

Vistanet. 2021. 'Lo sapevate? A Roma esiste un vero e proprio santuario dei gatti' [Did you know? There is a real cat sanctuary in Rome]. Accessed 13 December 2023. https://www.vistanet.it/roma/2021/02/05/lo-sapevate-a-roma-esiste-un-vero-e-proprio-santuario-dei-gatti/.

7
Feeding farm animals: perceptions and performances of the 'good farmer' among regenerative farmers

Hannah C. Mortimer

Introduction

The title of this book – *The Hand that Feeds* – harks back to the saying 'don't bite the hand that feeds you'. Human feeding of non-human animals (hereafter referred to simply as animals) is often done voluntarily and due to a desire to help animals (see Chapter 6 on stray cats and pigeons), which can have both positive and negative effects on their health and behaviour (see Chapters 8, 9, 10 and 11 in particular). However, it would be anthropocentric to claim that simply by feeding animals we are helping them, as this mentality also assumes that we know exactly what the animal needs to eat to be healthy and happy, and are able to control that. There are multiple instances where feeding animals is unhelpful because it leads to negative behaviours in those animals or makes them habituated to humans and reliant on them for food. In a farming context, it is perhaps more fit to talk about humans helping, or trying to help, animals (at least in the short term). This is especially the case when vets or farm workers administer medicines to sick livestock to help them recover (see Chapter 1 for a discussion of ritualised treatments for cattle in Roman antiquity). Farmers often argue that they were helping their animals in the sense that if they were not farmed and were wild or feral, they could die of sickness or starvation. Some of my interlocutors were involved in rearing native and rare-breed cattle, and argued that by farming them they were ensuring the continued existence of these breeds. These animals should also be viewed as 'workers' (Hamilton and Taylor 2013, 25) whose existence ensures the continuation of our pastoral landscape and Britain's food culture. After livestock are slaughtered, the human feeding of these animals is disrupted as this is when they now start the process of becoming food for us. In this sense, livestock are working for humans which involves them accessing food and shelter, and us accessing animal products. There is therefore an inherent contradiction between feeding and caring for animals and ensuring continuity of their breed on the one hand, and killing and eating them on the other. This tension makes livestock farming unique and an interesting context for exploring the complexities of care present in human–animal relationships.

As well as indirect feeding of animals (such as Chapters 4, 5 and 10), the types of feeding discussed in this volume also involve hand-feeding (such as veterinary treatments in Chapters 1 and 2). What is it about hand-feeding – that is, feeding an animal directly from one's hand – that is appealing? Dog trainers sometimes advise hand-feeding as it is said to have multiple benefits, including building a bond and trust with one's canine companion, especially if they are fearful or shy. Pets aside, perhaps there is something innately exhilarating about hand-feeding, especially a 'wild' animal. At several zoos and wildlife sanctuaries in the UK, one can opt to purchase this as an experience, such as at Wildwood in Devon, where one can choose an arctic fox, red squirrel or wolf 'feeding encounter' (Wildwood Trust Devon 2023). These animals are of course living in captivity rather than the wild, but the fact that they are often endangered or extinct in the local wilderness means that these experiences give us the opportunity to get up close and personal with creatures that we may never encounter in nature.

On livestock farms, hand-feeding of youngstock such as lambs, calves and kids with bottles of milk is a popular activity during spring, especially among young children. I can remember going to a 'lambing day' on a farm as a child and feeling excited about feeding and petting this small, woolly mammal. Despite hand-feeding being fun for adults and children alike (at least, those visiting farms), on many livestock farms, bottle-feeding milk to immature animals is done as a last resort. It is usually only necessary when the mother has died, is not producing enough colostrum (the first milk that humans and other mammals produce immediately after giving birth) or has rejected her baby and will not let it suckle. Thus, bottle-feeding sometimes becomes a necessity to keep the animal alive. Some farmers prefer not to bottle-feed as it creates a bond between human and animal, making it harder when the time comes for this animal to be slaughtered. It can also be a huge time commitment and inconvenience for farmers.

Despite the obvious connection we form in our minds between hands and feeding, many instances of animal feeding do not involve feeding directly from one's hands. This chapter draws on ethnographic fieldwork conducted between October 2022 and December 2023 on livestock farms in Devon that use regenerative practices, in which feeding the animals (cattle, sheep, pigs, goats and chickens) involves a mixture of rotational grazing, browsing and foraging in fields and eating a diverse diet including hay, silage, grains, grass and legumes. I was particularly interested in the care and consideration that went into feeding practices, but also how they related to farmers' sociocultural and environmental values and notions of 'good farming'. My methods were participant observation, interviews, 'go-alongs', walking interviews and visual methods. I used photography and video while in the field as an instant way of gathering data and as an *aide memoir* to my field notes, which I would write up later. I also used visual elicitation, whereby I asked my interlocutors to take photographs and videos of 'good farming', followed by an interview to discuss the images. I chose these methods to gain in-depth knowledge of perceptions and performances of good farming, especially in relation to feeding, in regenerative farming circles. My approach to pseudonymisation has been to choose names for the farms from English literature that I enjoy. My interlocutors' names have been changed to descriptive role-based pseudonyms to indicate their interests, personalities, experience or job role (for example, The Shepherd, The Cattle Farmer and so on).

The research that the chapters in this book are based on investigates the causes and consequences of animal feeding for human–animal–environment relationships, and focuses particularly on 'non-utilitarian' feeding (Thomas and Cassidy 2022, 2–3). Thomas and Cassidy define 'non-utilitarian' feeding as 'human feeding of other animals that is not part of a direct transaction or a means to an end' (2022, 2). The Introduction to this volume explains how feeding with utilitarian and effective aims involves 'feeding for the intentional generation of animal-derived products such as meat, eggs, milk, leather and traction', whereas non-utilitarian and affective feeding is 'emotional and non-product driven'. It might therefore seem counterintuitive to focus on farmers feeding their livestock. After all, aren't livestock being fed so that humans can make use of their bodies for our own gain? While this argument initially sounds credible, upon further reflection it is too simplistic, and a more nuanced understanding is required. The following paragraphs challenge the above argument by reflecting on what 'utilitarian' feeding means in the context of livestock farming.

First, although feeding farm animals is utilitarian on the one hand (they are being intentionally fed to generate animal-derived products), this is just one of livestock farmers' many goals. Other goals include increasing biodiversity, improving animal welfare and soil health, and sequestering carbon in the soil, to name just a few. Moreover, there are many affective (emotional) aspects to farmer–livestock relationships that complicate and disrupt the utilitarian/non-utilitarian binary (see Chapter 1). This chapter argues that regenerative farmers' perceptions and performances of 'good farming', especially their feeding practices, blur the boundary between what is utilitarian (transactional) and what is non-utilitarian (non-transactional and often affective). Food production – in this case, the production of meat products and associated by-products – on livestock farms is entwined with multiple dimensions of care. Moreover, as well as there being affective motivations for livestock farming (and hence feeding animals) regeneratively, there are also affective interactions between farm animals and farmers during feeding, whether that be an attachment created by bottle-feeding milk to lambs, affection in treating steers to some brewer's grains, or a sense of pride while watching one's cows frolic in the fields when they are let out of their barns and onto pasture in spring.

Having argued that generating animal-derived products is not the only purpose of farming, I want to return to the word 'utilitarian'. If, as the above definitions from Thomas and Cassidy (2022) and this book's Introduction clarify, utilitarian feeding is *intentional*, *product-driven*, and *transactional*, then any kind of animal feeding that is done *intentionally* has a utilitarian element, because there is a reason for doing it – a function, a purpose. Humans interact with animals for specific reasons even if they are unaware of them, and this is no different on a farm, at a zoo or in one's garden, because these animals essentially have a purpose or a function within the specific human–animal interaction of feeding. Other forms of animal feeding in the project (such as the feeding of captive animals in zoos, or wild birds in one's garden) can be seen as transactional in a broad sense because zoo visitors and people who feed wild animals are hoping to be entertained by the animals they see. (See Chapters 3, 8, 9 and 11 for the entertainment value of flea circuses, captive animals and tourism feeding.) In that sense, all forms of animal feeding are transactional, because we always get something back, even if we are gaining something less tangible than meat, milk, eggs, leather, wool, or traction, such as joy, satisfaction, fulfilment or belonging. As evident in Chapter 6 on the feeding of stray cats and pigeons, some humans voluntarily feed and care for animals because of an affective attachment to those animals and an ethic of care, which is linked to the desire to control or steward a species. Similarly, many farmers have an identity in which they are custodians of the countryside (Wheeler et al. 2018) and they affectively connect to the land (Baldwin et al. 2017). This involves a mixture of care for and control over ecosystems.

Feeding practices are generally important to research because these processes have consequences for humans, animals and the environment, including our collective health. This is especially true within livestock farming, whereby the animals' diets will also impact the health of our ecosystems and of the consumers. One need only think of the outbreak of Bovine Spongiform Encephalopathy (BSE) in 1986, and the human variant Creutzfeldt-Jakob disease (vCJD) to appreciate this. BSE was caused by feeding cattle (which are herbivores) mammalian meat and bone meal (MBM) and led to the deaths of millions of cows and some humans who ate beef from infected animals (Alarcon et al. 2023). Feeding practices on farms are also important to study because of their impact on the environment and sustainability, and finding ways to improve the sustainability of agriculture is paramount to tackling biodiversity loss and mitigating climate change (Kok et al. 2018; Smith et al. 2013).

Another reason for exploring affective motivations for and practices with livestock farming is that, as Bruckner et al. (2018, 37) argue, 'the literature on farm animals in both industrial and alternative agriculture has tended to neglect affective human–animal relationships … and replicated the nature–culture binary when addressing animal farming'. The nature–culture binary refers to the notion that there is a divide between nature and culture, biology and society, or humans and animals. In response to this long-held notion, the term 'naturecultures' was coined to argue that these things are entangled and dependent on each other in myriad ways (Haraway 2003). There are many affective practices within, and motivations for, livestock farming (such as caring for the animals, land and environment) which, as well as being related to food production, are also related to farmers' values, emotions, attitudes and beliefs. Human–livestock interactions are not purely instrumental (Wilkie 2010) and livestock can have ambiguous and unstable positions within farming systems, such as being pets on the one hand and a source of food on the other (Gorman 2017). People working within the livestock industry, from farmers to vets, pursue complex practices of care (Law 2010).

The 'good farmer' concept, which I will discuss in the following section, provides a useful analytical framework to explore these relationships because it is about farmers making value judgements, assessing each other's competence and forming an identity (Burton et al. 2020).

As animal feeding practices are integral to livestock farming, it seems odd that they are not discussed much in social scientific literature. Farmers' decisions on what and how to feed their livestock are important ones that impact their social and cultural capital. Extant literature discusses the environmental aspects of the good farmer, notably how farmers' engagement with agri-environment schemes (AES) reflect productivist good farmer identities (Burton 2004; Burton et al. 2008), but within these analyses, animals have been left out of the picture. More recently, a growing body of literature is exploring the way farmers (and others who work with animals to produce food, such as abattoir workers and meat inspectors) engage with animals, often in conflicting ways (Wilkie 2005, 2010; Hamilton and McCabe 2016; Bruckner et al. 2018; McLoughlin and Casey 2022). Expanding on these works, I have chosen to focus on human–animal relationships within regenerative agriculture – a social movement, set of practices and approach to farming that aims to generate more beneficial environmental impacts (Newton et al. 2020; Knepp Castle Estate n.d.). Potential outcomes of regenerative agriculture include greater biodiversity, flood protection, improved soil health and carbon sequestration (Newton et al. 2020). Regenerative agriculture takes a holistic approach to farming, viewing the environment, animals and people as being connected to benefit landscapes and ecosystems.

This chapter begins by outlining the 'good farmer' concept and its relevance to human–animal relationships. This is followed by a discussion on regenerative agriculture and a summary of feeding practices on regenerative farms, with a focus on farms predominantly rearing beef cattle and sheep. Using data collected during ethnographic fieldwork, I argue that good farming is a useful lens for exploring the role of food, feeding and eating in regenerative systems because it involves visible performances from farmers. Regenerative farmers' performances of good farming involve controlling visible symbols such as grazing livestock, cover crops, herbal leys and biodiversity, the first three being directly related to feeding animals. The way farmers control these symbols or perform their tasks is sensory, with vision, hearing and touch used to mob-graze cattle, for instance. My interlocutors also talked about production alongside multiple caring actions, including improving animal welfare, biodiversity and soil health, revealing the affective motivations for farming and aspects of the good farmer identity that are about more than merely production.

What makes a 'good farmer'?

Although a variety of research has been conducted on the 'good farmer', conceptions are varied and there is no overarching theory. On the contrary, the good farmer means different things to different people, including scientists, policymakers, academics and farmers themselves (Burton et al. 2020). It is 'an established concept used to identify which practices are respected by members of the farming community' (Cusworth 2020, 164). Burton et al. (2020, 14) argue that the term is used as both a 'common-sense category', in that farmers use it to judge each other's practices and competencies, and an 'analytical construct', in that academics and policymakers use it to suggest improvements to farming practices or to try to understand farming culture. However, 'the essential meaning remains the same – the "good farmer" term is used to refer to the extent of cultural competency/morality in farming either by those within the peer group, those seeking to influence the peer group, or those studying the cultural construction and functioning of the peer group' (Burton et al. 2020, 8). Within my research, I use the conceptualisations from both farmers and social scientists.

According to Burton et al. (2020), during the eighteenth and nineteenth centuries good-farmer ideals shifted from ideas of identity, morality, family and community to perfecting one's productivist skillsets. They argue that 'through this process, moral and cultural aspects of modern farming have been hidden under a veil of scientific rationality' (Burton et al. 2020, 64). Productivism refers to new policies and approaches to agriculture that accelerated after the Second World War and that aimed to modernise agriculture and increase productivity and output through more intensive, industrial and expansive systems (Lowe et al. 1993, 221). Productivism is an ideology and a set of farming practices and policies with unintended detrimental impacts on the environment (Wilson 2001), such as increased biochemical inputs, leading to pollution, as well as intensification, leading to loss of biodiversity and soil pollution, compaction and erosion (Wilson, 2007, 94–5). Despite increases in productivist mindsets, changes in the 'rules of the game' have caused many farmers to change their behaviour, values and ideas of good farming (Sutherland and Darnhofer 2012, 232). I am particularly interested in the affective aspects of the good-farmer identity, which are not just about production, but also relate to how farmers care for their animals and the land, how they negotiate the contradictory notions of care within livestock farming, or how they perform their roles in front of each other and the public.

These affective aspects, although aiding farmers' primary focus of producing quality food, are also related to their values, emotions and beliefs and are not always 'in the name of production' (Burton et al. 2020, 64). Furthermore, there is a great diversity of farming systems, and the people who work within them have different ideas of what constitutes a good farmer.

Conceptualisations of 'good', and by extension 'bad', farming among my interlocutors were gradually revealed during my ethnography. Bad farming was mostly thought to include practices and attitudes that neglected or degraded the environment:

> I think bad farming is farming whereby the farmer doesn't really have an interest in the land, he's just using the land. He might be an insurance investor, he might be sitting at a desk in London, he might be a builder, he or she might be a Lord or Lady who doesn't really visit the farm. They've got vast acreages. And they are just telling contractors what to do, to spray a thousand acres of this and do your normal thing. There is no connection there, it's just a way of making money. It doesn't sit comfortably with me. It's a responsible thing. (The Cattle Farmer, regenerative beef farmer)

> I'd say any farm that degrades the environment, I would classify as bad farming. I wouldn't say they are bad farmers, but it's bad farming as in, there are other ways we can do it. (The Apprentice, trainee regenerative farmer)

What is regenerative agriculture?

Researchers such as Burns (2021, 54) have described regenerative agriculture as a 'social movement' and 'paradigm shift' within farming, which views land, water and nutrients 'as an ecological whole'. Burns explains that 'regenerative agriculture is a proposal about changing farming in order to undo the degradation of the farmed environment. It is a shift towards farming *with* the environment, rather than treating it as merely a platform' (Burns 2021, 55). Hence, regenerative agriculture is arguably related to the affective aspects of the good-farmer concept that are not merely production oriented. However, regenerative agriculture is a broad and flexible term; definitions vary among both scholars and practitioners (Newton et al. 2020). The Manager summed this up nicely:

> Regenerative agriculture is very much a buzzword at the moment … it has quite a broad definition, I feel. And there are different sorts of interpretations of what exactly regenerative agriculture is, according to what people are trying to prove or sell and also which kind of sector of agriculture you're looking at. To arable farmers, it's about minimum tillage. To organic arable farmers who have no other way of controlling weeds, minimum tillage is really difficult to achieve. But they don't use glyphosate [a herbicide], so, you know, you've got that whole argument there. So with livestock farming and regenerative, I think it's just a more holistic approach and less focused on output. (The Manager, regenerative livestock farmer)

When asked to define regenerative agriculture, regenerative farmers mentioned improving the land but also being open to trying different things due to the variety of farming landscapes and systems:

> Sustainable is not enough, because we have degraded the land so much that why would you want to sustain it? So sustainable is not enough, we need to regenerate it back to its full potential … And if you wanted to complicate further, you could add in using the six principles. I think Gabe Brown said five, but someone else added in a sixth which is context. I think it makes a lot of sense because it allows for adaptability for different farms and things, which I think a lot of people don't realise when they set out on it. They might think that any size fits all but it's definitely not the case. (The Entrepreneur, regenerative pig and poultry farmer)

Indeed, practitioners and advocates of regenerative agriculture generally agree that it should be based on the following five principles: limit disturbance of the topsoil, cover the soil with crops all year round, plant a diversity of crops, retain living roots, and integrate livestock (Cherry n.d.; Gosnell 2022; Gosnell et al. 2019; Hintz 2015; Knepp Castle Estate n.d.; Newton et al. 2020; SRUC n.d.; White 2020). Integrating livestock often involves rotational grazing. This entails dividing up one's fields into smaller paddocks and regularly rotating the livestock between them so that the pasture has time to recover (Department for Environment, Food and Rural Affairs 2021). Rotational grazing can include 'mob grazing' where livestock intensively graze one paddock for a short time and are moved at least once a day, or 'forward grazing', in which two groups of livestock, usually from the same species but with different nutritional needs, take turns to graze each paddock. The UK Government claims that rotational grazing increases forage, improves soil health and fertility and reduces pollution and feed costs (Department for Environment, Food and Rural Affairs 2021). Other alleged benefits are improved weed control and greater biodiversity (Soil Association Scotland n.d.; SRUC n.d.). Rotational grazing aims to extend the grazing period, thus reducing the feed costs that would have gone into housing the cattle for this period (AHDB 2023).

Proponents of rotational grazing argue that by regularly rotating livestock around fields one is mimicking the movements made by large herbivores (such as aurochs, bison and buffalo) thousands of years ago and before the domestication of cattle. These animals would constantly be on the move, searching for food and escaping predators (White 2020; Morris 2021):

> So what makes it good is that you are acting as a predator. [Today's cattle] have no natural predators [although their ancestors would have]. They are herd animals, so in the wild they would consistently move on. So they're not in one spot for too long so they don't trample the ground, cause compaction or damage the soil and microecology. So the idea is the grass doesn't get fully grazed down, hooves don't cause too much damage, so it recovers quicker and you don't damage those microorganisms. (The Apprentice, trainee regenerative farmer)

Cattle dung, urine, and plants, which are trampled into the ground by their hooves, are seen as beneficial as they help to fertilise and store organic matter in the soil, which also improves soil carbon (Morris 2021). Some proponents of extensive and rotational grazing also argue that keeping livestock outside for longer periods of time by extending the grazing season is beneficial for the climate and biodiversity, due to livestock being part of the biogenic carbon cycle. This cycle involves the recycling of carbon: ruminants release carbon (C) as methane (CH_4) through their belches, but this is converted into carbon dioxide (CO_2) after about twelve years through a process called hydroxyl oxidation (Werth 2020). Carbon dioxide is then absorbed by plants, which release oxygen (O_2) through the process of photosynthesis (Werth 2020). Carbon is then stored as carbohydrates in plants, which can be eaten by ruminants, beginning the cycle again (Werth 2020). By utilising cover crops and retaining living roots, more carbon can be sequestered in the soil (White 2020). However, claims that grazing ruminants can help to mitigate climate change through sequestering carbon in soils, and that intensive, rotational grazing is better than conventional set stocking (that is, keeping a group of cattle in a field until they have eaten everything) for improving productivity and biodiversity, remains debatable (Morris 2021). In fact, due to the specific ecosystems of individual farms, rotational grazing may not be successful everywhere.

Regenerative agriculture is generally seen as a positive movement, having been embraced by some companies, including Waitrose and Nestlé, and is growing in popularity, especially among new entrant farmers. However, it has also received some criticism. *The Guardian* columnist and activist George Monbiot argues that organic, extensively raised ruminant livestock are actually *more* damaging for the environment due to the vast amounts of land they use vis-à-vis the small percentage of protein they contribute towards people's diets (Monbiot 2022). Extensive ruminant livestock systems require more land than intensive ones, while monogastrics (pigs and poultry) require even less land because they have better feed-conversion efficiency (Garnett 2010, 3). However, making comparisons between the land use and feed-conversion efficiency of different systems alone is too simplistic. There are multiple other factors to consider in these debates, including the benefits to animal welfare (Grandin 2022), biodiversity (Teague and Kreuter 2020) and human health (Butler et al. 2021) from extensive ruminant or pasture-based systems, as well as supporting people's livelihoods. My ongoing research aim is not to assess the validity of the varied claims made about livestock farming, regenerative agriculture, and intensive versus extensive systems, but to explore the affective side of human–animal relations on regenerative farms and how they relate to the concept of the 'good farmer'.

As regenerative agriculture emphasises the importance of rotational grazing, many regenerative farmers also bear the 'Pasture for Life' certification that comes from being a member of the Pasture-Fed Livestock Association (PFLA). Many of these farms are also organic. To be a PFLA farmer, one must commit to feeding one's livestock only pasture and forage – except for milk, which is drunk by youngstock before weaning. All other types of feed, including grains and soya, are prohibited (Pasture for Life 2021, 20). Two farms that I worked with were Pasture for Life certified and one of the farmers is especially proud of this, selling his PFL produce in his own farm shop.

Feeding practices on regenerative livestock farms

There is a huge diversity of feeding practices used on regenerative livestock farms in the UK. Within many beef systems in the UK, cattle are usually housed in barns for part of the year and fed on hay or silage (preserved forage) during autumn and winter (Animal Welfare Committee 2021). However, within regenerative agriculture there is a movement to extend the grazing season by using bale grazing (unrolling bales of hay in fields for cattle to eat) in the autumn and even outwintering cattle so that they spend their whole lives outdoors. Regardless of the system, sheep are kept outside and are grazing most of the year but are often brought inside for lambing. Within many systems, beef cattle and sheep diets tend to be supplemented with grain such as wheat or barley, legumes such as soya, or small pellets (a mixture of concentrates called 'cake'). Although conventional farming is far more common, the fact that pesticides and herbicides are permitted, and the carefully controlled measurements that are required, encourages some farmers to adopt organic practices. The Cattle Farmer reflected on his transition to organic and then regenerative practices:

> [Conventional farming] was very controlling. That's what I was questioning I think, through those early years. And now I realise it was a big con! That's with the dairy and arable. And with the beef and sheep at college times it was a bit more relaxed but still prescriptive, such as lambing and feeding ewes at a specific time ... I guess in a way they had to do that because it was for the farmers' benefit, but I don't know, I used to cringe at it. Anyway, I wanted to take more interest in organic farming ... And since 1992 we have been organic and I have been trying to see how organic we can go, such as now we are doing Pasture for Life, so we are raising our cattle on 100 per cent pasture, no cereals at all. We are exploring mob grazing, regenerative farming, trying to build a more resilient farm in terms of weather extremes. (The Cattle Farmer, regenerative beef farmer)

As this individual suggests from his mention of Pasture for Life, organic regenerative systems rely on there being enough pasture. This is predominantly grass, but can include species-rich permanent pasture (land that has been used for more than five years to grow herbaceous forage, which can be sown or self-seeded) and herbal leys (temporary grasslands including various legumes, herbs and grasses). Pasture and preserved forage form the bulk of cattle and sheep diets, and less grains or concentrates are used. This process is seen by these farmers as being both more environmentally friendly and better for animal welfare, with grazing being used to improve biodiversity but also being viewed as a natural ruminant behaviour. Ruminants are herbivorous animals (including sheep, cattle, goats and deer) that chew the cud (chew partially digested food) which has been fermenting in their rumen (stomach). Although the principles of regenerative agriculture are well documented, their impacts on human–animal relationships are not. There is some literature, especially from sociology and geography, on human–livestock relationships in conventional farming systems (Wilkie 2010; Ellis 2013), but fewer studies have explored more-than-human relations and care within the context of regenerative agriculture. Kallio and LaFleur (2023) have engaged with these topics and related them to feeding: they explore more-than-human relations through ethnographic research on regenerative farms in Finland, Norway and Italy. They argue that livestock had various roles on the farms in addition to producing meat for human consumption, such as providing by-products (wool, eggshells) that were then re-circulated both within and outside the farms: wool could be sold to produce textiles but also provided a plant mulch on the farms, while eggshells were composted to improve the soils (Kallio and LaFleur 2023, 9). Interdisciplinary research within the social sciences and environmental humanities has also discussed human–soil relations and care within the context of regenerative agriculture (Krzywoszynska 2019).

What links the 'good farmer', regenerative agriculture and feeding?

If being a good farmer involves developing social and cultural capital by having visible symbols of production (Burton 2004; Burton et al. 2008), evidence of environmental land management (Wheeler et al. 2018) and conservation activities (Lavoie and Wardropper 2021), then feeding practices are symbols of good farming as they relate to these aspects of the good-farmer identity. Feeding practices incorporating grazing are highly visible to other farmers and passers-by, and being a good farmer requires one's practices and productivity to not just be visible, but also aesthetically and morally pleasing, whether that be, for example, tidy fields and straight lines (Burton 2004; Burton et al. 2020), healthy livestock (Naylor et al. 2016; Vigors et al. 2023), or conservation tillage (Lavoie and Wardropper 2021). The physical condition of one's grazing livestock is a visible aspect of the landscape that can be observed over hedgerows by other farmers (Burton 2004; Overstreet 2018). The farmers in Burton's study referred to this practice as 'hedgerow or roadside farming', 'which involves a two-way interaction between the displayer … and the observer' (Burton 2004, 204). Interestingly, this practice was 'a status symbol in itself' – farmers observed and expected to be observed, as this was part of being a good farmer (Burton 2004). If farm animals are not fed well, it is evident from their appearance, impacting the status of their owner due to the perceived lack of skill (Overstreet 2018). Wilkie (2010, 41) also argues that livestock are a visual manifestation of 'good stockmanship', symbolising farmers' skills, reputation and identity.

Some farmers use specific grazing practices because they perceive them to be better than other practices (whether for environmental, animal welfare or economic reasons) or because they want to be seen as good farmers by other people, namely their peers and stakeholders within the farming community. For example, mixed farmer and author James Rebanks (2020a) argues that conserving one's farmland should involve rotational grazing and mixed rotational farming. Mixed farming refers to a farming system that includes livestock and arable crops on the same farm (Grigg 1974); mixed rotational farming involves rotating one's crops and livestock, which is said to have many benefits such as improving soil health and controlling weeds, potentially increasing productivity and sustainability (AHDB 2022).

For some English farmers, pressures from the government to produce multiple 'public goods' on their farms as well as food (such as reduced greenhouse gas emissions, increased biodiversity, access to nature etc.; Cusworth and Dodsworth 2021) mean that attitudes are changing, and mismanaging the environment can now negatively impact one's social and cultural capital (Cusworth 2020). The fact that, for some people, the definition of the good farmer is changing is another reason to focus on regenerative agriculture, which strives to create positive changes to the environment. While there has been growing awareness of the environmental impacts of intensive agriculture (such as biodiversity loss and greenhouse gas emissions) and especially of ruminant livestock production (Cusworth et al. 2021), advocates of regenerative agriculture argue that, if managed correctly, ruminant livestock can be beneficial for the environment and even reverse previous damage by restoring soil health and enhancing ecosystem services (Teague and Kreuter 2020).

Although ideas of good farming among farmers have shifted in recent years towards more environmentally friendly practices (such as using less pesticides, herbicides and synthetic fertilisers, reducing tillage or changing one's animal feed), productivist narratives are common (Lavoie and Wardropper 2021). But productivity and yield are now more entangled with environmental objectives such as increasing biodiversity, or cultural morals such as improving animal welfare, and agricultural workers are often thought to have an 'ethics of care' (Cusworth 2023).

Using one's senses to care for livestock

Farmers use sensory skills and embodied knowledge on a daily basis (Grasseni 2004; Burton et al. 2008; Carolan 2008), which can also involve expressing emotions. Grasseni (2004, 41) has developed the notion of 'skilled vision'. She theorises how dairy cattle farmers use an embodied, sensorial practice of skilled vision as a way of understanding and knowing about their Alpine Brown cows. Skilled vision is 'the way our looking can be trained to detect certain specific features in objects that are commonly available to generalised perception' – the object of study here being cows (Grasseni 2004, 49). Crucially, this skill is a learned practice through years of training and experience in breeding cattle, and needs constant nurturing and development through a form of 'apprenticeship' (Grasseni 2004, 45). Despite this, 'skilled vision can be intrinsic to [cattle breeders'] identity' (Grasseni 2004, 42). Furthermore, skilled vision involves a combination of looking and touching to ensure the animal is healthy and adheres to certain physical requirements. Grasseni (2004) implies that her interlocutors' approach to breeding can be both utilitarian and affective, as although they have practical and financial interests in which cows should be bred and sold in cattle fairs, their decisions are also dependent on romantic ideals such as notions of beauty and heritage. How farmers breed and care for their cattle matters, as different people have different moral codes about what physical characteristics make the perfect-looking dairy cow. Moreover, Grasseni highlights how farming practices, characterised by an embodied attentiveness to the needs and condition of one's cattle, are at odds with the 'panoptic', 'disembodied gaze' present at cattle fairs in which cows and their body parts are seen for their usefulness (Grasseni 2004, 51). Despite this tension, the idea of beautiful cows is also affective. Overall, Grasseni is interested in how processes of learning skills while educating and honing one's senses can create a shared vision that is embodied among cattle breeders, which she too was eventually able to enact after working as their apprentice.

As preserved forage and other livestock feed are produced and manufactured by humans and then directly given to the animals, farmers use skilled vision during these interactions. During the autumn and winter months of my fieldwork, I witnessed farmers using vision, smell and touch, especially when distinguishing between straw and hay. For instance, The Cattle Farmer is able to tell the difference between straw and hay by plucking out a few pieces, looking at them to see which plants they came from, and feeling their texture in his hands. To someone from a non-farming background such as myself, it was initially difficult to notice the difference, yet it is important that farmers can distinguish between straw and hay as they have different purposes. Straw is usually used for 'bedding up' because it is less nutritious, whilst hay is used for feeding during the colder months of the year and as a supplement when there is a lack of forage (such as during dry summers). Straw mostly contains the stems of grain plants such as oat, barley or wheat, whereas hay is dried grass. Silage is also preserved grass, but it has been wrapped in plastic to start a process of fermentation, whereas hay and straw are baled and left to dry in the sun.

The Show-goer (a beef and sheep farmer and regular attendee of agricultural shows) used her sense of taste while chewing on a piece of grass from the hay that was being fed to her cattle, to test how sweet it was. She and her husband The Hobby Farmer also feed their cattle brewers' grains as a treat, which are by-products left over from the family brewery. They are a good source of protein and apparently the cattle 'love it':

> **The Show-goer**: When [our son] is brewing, when he is boiling up the mash, you can smell the wort, the barley, has got that sweet smell in the air, and the cattle know, they stand there at the gate smelling it.
>
> **The Hobby Farmer**: When we start brewing the cows start mooing!
>
> **The Show-goer**: And so [our son] takes it out to them in the fields and they love it, they go wild for it. They love eating it. And often they get it when it is still slightly warm, which I think must be extra nice for them. And it is the only grain that they get fed. They are basically grass-fed but we allow ourselves to make an exception with the brewers' grains.
>
> **The Hobby Farmer**: We take the sugar out of it to make the beer, but you have still got all the cellulose and whatever else is in grain, so it has nutritional value.

This farming couple enjoy feeding their cattle brewers' grains because otherwise these by-products would be wasted. They are thus creating more of a closed loop system in which nutrients and organic matter are recirculated within the farm. The above quotations demonstrate the affective dimension of human–animal interactions, because The Show-goer and The Hobby Farmer have other objectives as well as the utilitarian one of rearing cattle for beef: their description focuses on how the cattle react and appear to enjoy the treat, so they are thinking about what their cattle *like* eating (and they clearly enjoy feeding them). This is a multi-species sensory experience, with the smell and heat from the mashed-up grains, and the sound of the cattle bellowing providing an insight into farmer–livestock interactions. The interaction between a farmer and his cows can be 'satisfying' and 'emotional' at the same time, as expressed by The Hobby Farmer:

> One of the most satisfying things is after the tractor has finished, and it's dark and you go down there with a torch, and you stand there on the feed side of the barrier, and they are all eating, and there is the wonderful smell of the haylage, and there is this contented munching. All these cows which might usually keep their distance from each other and might have a fight over who is the top cow, they are all cheek by jowl eating. And I think there is an emotional exchange between them and you, because they know that you fed them because they saw you doing it, you smell of the stuff because you have been cutting the things up with your hands, and they let you get near to them, you can stroke their heads and all the rest of it. (The Hobby Farmer, regenerative beef and sheep farmer)

Throughout the year, regenerative farmers rely heavily on their senses of vision and hearing to feed their cattle. As part of my participant observation on farms, I have been involved in the regular moving of cattle across fields. The following vignettes describe my observations while moving the cattle on two different occasions.

> *11th November 2022*
> Today I took part in a 'drive along' and moved the cattle in Hobbiton. Along with my interlocutors in a trailer, we drove along a bumpy road from the farm headquarters to another field. Small birds swiftly flew away as we hurtled past hedgerows and ditches. Our job was to move the cattle from one field to another field

where they could graze fresh pasture. Whilst waiting for instructions from The Cattle Farmer on when to start moving the cattle, The Shepherd noticed an injured thrush hopping along on one leg. She tried to catch it as she said she wanted to take it home and look after it, but sadly she couldn't get hold of it as it fearfully darted into a bush. My job during the cattle drive was to stop traffic whilst the cows were moving past, allow the traffic to pass if it was safe to do so and stop the animals from walking the wrong way. I did the latter by spreading my arms wide and holding a piece of plastic piping in one hand to make myself bigger. We were all wearing hi-vis jackets. I found out later that a lot of shouting and gesticulating was also required. 'Come on! Come onnnnnnn!' shouted the farmers. They were trying to attract the attention of the cows and make them follow the sound of their voice. The cows seemed interested in me and started walking towards me (not what they were supposed to do) so The Cattle Farmer instructed me to wave my arms in big circles and shout 'Shoo!'. If I made myself bigger, they would turn around and walk towards the 'gap' – that is, the part of the road that was not blocked by a person or vehicle. I did as I was told but felt rather silly and self-conscious, worried that I might be doing it wrong as the cows got increasingly closer. I was told later by The Shepherd that I did well, but The Traveller said that I should try to be louder next time. The Shepherd did admit that a lot of people are not used to shouting.

7th February 2023
This morning I helped The Cattle Farmer with mob grazing the cows that were brought outside last week due to the outbreak of pneumonia amongst the herd. I struggled to keep up with him as he marched up the hill. It was a cold, crisp morning and the steep incline left me panting for breath. Birdsong resounded all around us. A layer of mist blanketed the valley, and the sky was a bright blue; dewy grass glistened in the morning sunshine. 'Morning cows!' The Cattle Farmer called, as I took a photo [Figure 7.1]. One of them answered with a loud bellow. I followed The Cattle Farmer and helped him move the electric fence, creating an opening for the cows to walk through. He called to them 'Come on!' and they started running through the opening to the new section. I took a video whilst this happened – the difference in the condition of the sward between the paddock they were in and the new one is very obvious. In the old paddock, the grass has been grazed tighter and

Figure 7.1 The Cattle Farmer walks through his herd of Hereford cattle to set up a new paddock for them to graze. *Source*: author, with permission from farmer.

there are more yellow stems and bare patches of ground where the animals have trampled it, whereas the grass in the new paddock is thicker and lusher. The cows were clearly pleased to have new grass to eat, as they rushed through the opening and immediately started grazing.

These vignettes show how farmers use their sense of vision and hearing in the practice of mob grazing. Farmers need to be alert, able to see their cattle and what they are doing, as well as have good hearing, the ability to project their voices and understand cattle behaviour. These requirements form some of livestock farmers' unique skillsets. Moving cattle is a sensory experience, as they can bellow loudly when they are hungry or frustrated (they do not always appreciate being moved). For some animals, the excitement of having fresh grass to eat is evident in their playfulness as they enter a new field or paddock (they will often run down hills for no obvious reason). As mentioned in the introduction to this chapter, this playfulness is especially evident in spring when they are let out of barns they have spent the winter in, and led onto pasture. In the first vignette, I was performing the role of a good farmer or good stockperson alongside my interlocutors – by communicating with the cattle through the sound of our voices, we were showing the locals and others driving by that we were communicating with the animals and moving them to a new location, which requires skill and patience. I experienced first-hand what it feels like to be noticed by passers-by, who may have assumed that I was a farmer too and were judging my competency at herding cattle.

Conclusion

This chapter reflected on how regenerative farmers' values, attitudes and beliefs about good practices impact the way they feed and care for their animals. I began this chapter by reflecting on the title of this book and questioning whether, by feeding, we are always helping or providing a service to animals, or whether we are doing it for our own ends. Whether it is farmers feeding their livestock (Chapter 1), people feeding captive animals (Chapters 2, 3, 8, 9, 11), or people feeding wildlife (Chapters 4, 5, 6, 10), how much are we benefiting animals, and should we do anything differently? Although there is no scope to answer these questions in this chapter, I encourage readers to reflect. The good-farmer concept remains a useful lens for discussing the role of food and feeding, because it involves making value judgements about the best way to feed one's livestock, and because it involves visible performances from farmers that enhance their social and cultural capital.

British concerns over environmentalism, conservation and animal welfare, as well as changes in agricultural legislation and policies, have shaped how farmers work with their animals and the environment, which inevitably includes thinking about how to improve one's feeding practices. As The Manager stated, regenerative agriculture is now a 'buzzword', and so, despite it being a broad and flexible term, farmers and food retailers are increasingly interested in the movement, marketing themselves as regenerative or transitioning to regenerative procurement or practices. However, The Manager also claimed that on her farm, they were practising regenerative agriculture before the term became popular. I am reminded of something that was said at Rootstock, a conference I attended in Exeter in February 2023, which I found particularly resonant. One of the organisers of Groundswell – an annual event to educate on regenerative agriculture – said that it involved 'old wisdom that still works'. He explained that regenerative agriculture is not a new concept and many of the things that farmers are learning to do now were done at least a century ago. Unfortunately, this old wisdom was gradually forgotten by many farmers due to the way they were educated and trained after the Second World War, encouraged by several successive governments to focus on increasing production at all costs (Brassley et al. 2021; Rebanks 2020b). What is now left to explore is how regenerative farmers negotiate complex and contradictory requirements, such as caring and killing or conserving and producing. How do regenerative practices impact human–animal relationships? These questions are too broad to be answered now, but they should be considered by scholars, for they are new and exciting areas of research.

References

AHDB. 2022. 'Livestock and the arable rotation'. Accessed 16 March 2022. https://ahdb.org.uk/livestock-and-the-arable-rotation.

AHDB. 2023. 'Rotational grazing systems for cattle'. Accessed 17 April 2023. https://ahdb.org.uk/knowledge-library/rotational-grazing-systems-for-cattle.

Alarcon, P., B. Wall, K. Barnes, M. Arnold, B. Rajanayagam and J. Guitian. 2023. 'Classical BSE in Great Britain: Review of its epidemic, risk factors, policy and impact', *Food Control* 146 (109490): 1–17.

Animal Welfare Committee. 2021. *AWC Opinion on the Welfare of Cattle Kept in Different Production Systems*. London: Department for Environment, Food and Rural Affairs.

Baldwin, Claudia, Tanzi Smith and Chris Jacobson. 2017. 'Love of the land: Social–ecological connectivity of rural landholders', *Journal of Rural Studies* 51: 37–52.

Brassley, Paul, David Harvey, Matt Lobley and Michael Winter. 2021. *The Real Agricultural Revolution: The transformation of English farming, 1939–1985*. Woodbridge: Boydell Press.

Bruckner, Heide K., Annalisa Colombino and Ulrich Ermann. 2018. 'Naturecultures and the affective (dis)entanglements of happy meat', *Agriculture and Human Values* 36: 35–47.

Burns, Edgar A. 2021. 'Regenerative agriculture: Farmer motivation, environment and climate improvement', *Policy Quarterly* 17 (3): 54–60.

Burton, Rob J. F. 2004. 'Seeing through the "good farmer's" eyes: Towards developing an understanding of the social symbolic value of "productivist" behaviour', *Sociologia Ruralis* 44 (2): 195–215.

Burton, Rob J. F., Jérémie Forney, Paul Stock and Lee-Ann Sutherland. 2020. *The Good Farmer: Culture and identity in food and agriculture*. London: Routledge.

Burton, Rob J. F., Carmen Kuczera and Gerald Schwarz. 2008. 'Exploring farmers' cultural resistance to voluntary agri-environmental schemes', *Sociologia Ruralis* 48 (1): 16–37.

Butler, Gillian, Ali Mohamed Ali, Samson Oladokun, Juan Wang and Hannah Davis. 2021. 'Forage-fed cattle point the way forward for beef?', *Future Foods* 3 (100012): 1–7.

Carolan, Michael S. 2008. 'More-than-representational knowledge/s of the countryside: How we think as bodies', *Sociologia Ruralis* 48 (4): 408–22.

Cherry, John. n.d. '5 principles of regenerative agriculture'. Accessed 10 May 2022. https://groundswellag.com/principles-of-regenerative-agriculture/.

Cusworth, George. 2020. 'Falling short of being the "good farmer": Losses of social and cultural capital incurred through environmental mismanagement, and the long-term impacts of agri-environment scheme participation', *Journal of Rural Studies* 75: 164–73.

Cusworth, George. 2023. 'Metabolic agricultural ethics: Violence and care beyond the gate', *Progress in Environmental Geography* 2 (1–2): 58–76.

Cusworth, George and Jennifer Dodsworth. 2021. 'Using the "good farmer" concept to explore agricultural attitudes to the provision of public goods: A case study of participants in an English agri-environment scheme', *Agriculture and Human Values* 38: 929–41.

Cusworth, George, Tara Garnett and Jamie Lorimer. 2021. 'Agroecological break out: Legumes, crop diversification and the regenerative futures of UK agriculture', *Journal of Rural Studies* 88: 126–37.

Department for Environment, Food and Rural Affairs. 2021. 'Use rotational grazing on permanent grassland'. Accessed 28 February 2022. https://www.gov.uk/guidance/use-rotational-grazing-on-permanent-grassland.

Ellis, Colter. 2013. 'The symbiotic ideology: Stewardship, husbandry, and dominion in beef production', *Rural Sociology* 78 (4): 429–49.

Garnett, Tara. 2010. 'Intensive versus extensive livestock systems and greenhouse gas emissions'. Accessed 18 June 2024. https://tabledebates.org/sites/default/files/2020-10/FCRN_int_vs_ext_livestock.pdf.

Gorman, Richard. 2017. 'Therapeutic landscapes and non-human animals: The roles and contested positions of animals within care farming assemblages', *Social & Cultural Geography* 18 (3): 315–35.

Gosnell, Hannah. 2022. 'Regenerating soil, regenerating soul: An integral approach to understanding agricultural transformation', *Sustainability Science* 17 (2): 603–20.

Gosnell, Hannah, Nicholas Gill and Michelle Voyer. 2019. 'Transformational adaptation on the farm: Processes of change and persistence in transitions to "climate-smart" regenerative agriculture', *Global Environmental Change* 59 (101965): 1–13.

Grandin, Temple. 2022. 'Grazing cattle, sheep, and goats are important parts of a sustainable agricultural future', *Animals* 12 (16): 1–13.

Grasseni, Cristina. 2004. 'Skilled vision: An apprenticeship in breeding aesthetics', *Social Anthropology* 12 (1): 41–55.

Grigg, David B. 1974. *The Agricultural Systems of the World: An evolutionary approach*. Cambridge: Cambridge University Press.

Hamilton, Lindsay and Darren McCabe. 2016. '"It's just a job": Understanding emotion work, de-animalization and the compartmentalization of organized animal slaughter', *Organization* 23 (3): 330–50.

Hamilton, Lindsay and Nik Taylor. 2013. *Animals at Work: Identity, politics and culture in work with animals*. Leiden: Brill.

Haraway, Donna J. 2003. *The Companion Species Manifesto: Dogs, people and significant otherness*. Chicago: Prickly Paradigm Press.

Hintz, C. M. H. 2015. '"Soil in My Blood": Women farmers, transformative learning, and regenerative agriculture', doctoral thesis, Prescott College. Accessed 23 May 2022. https://www.proquest.com/docview/1694872242/abstract/CDC3CB434B6D441FPQ/1.

Kallio, Galina and Will LaFleur. 2023. 'Ways of (un)knowing landscapes: Tracing more-than-human relations in regenerative agriculture', *Journal of Rural Studies* 101 (103059): 1–13.

Knepp Castle Estate. n.d. 'Regenerative farming at Knepp'. Accessed 5 May 2022. https://www.kneppestate.co.uk/regenerative-agriculture.

Kok, M. T. J. and 15 others. 2018. 'Pathways for agriculture and forestry to contribute to terrestrial biodiversity conservation: A global scenario-study', *Biological Conservation* 221: 137–50.

Krzywoszynska, Anna. 2019. 'Caring for soil life in the Anthropocene: The role of attentiveness in more-than-human ethics', *Transactions of the Institute of British Geographers* 44 (4): 661–75.

Larder, Nicolette. 2021. 'Good farming as surviving well in rural Australia', *Journal of Rural Studies* 88: 149–56.

Lavoie, Avery and Chloe B. Wardropper. 2021. 'Engagement with conservation tillage shaped by "good farmer" identity', *Agriculture and Human Values* 38: 975–85.

Law, John. 2010. 'Care and killing tensions in veterinary practice'. In *Care in Practice: On tinkering in clinics, homes and farms*, edited by Annemarie Mol, Ingunn Moser and Jeannette Pols, 57–71. Bielefeld: transcript Verlag.

Lowe, Phillip, Jonathan Murdoch, Terry Marsden, Richard Munton and Andrew Flynn. 1993. 'Regulating the new rural spaces: The uneven development of land', *Journal of Rural Studies* 9 (3): 205–22.

McLoughlin, Eimear and John Casey. 2022. 'On "finishing": A visual memoir of care and death on an Irish cattle farm', *Visual Anthropology Review* 38 (1): 34–59.

Monbiot, George. 2022. 'The most damaging farm products? Organic, pasture-fed beef and lamb'. *The Guardian*. Accessed 12 June 2024. https://www.theguardian.com/environment/2022/aug/16/most-damaging-farm-products-organic-pasture-fed-beef-lamb.

Morris, C. D. 2021. 'How biodiversity-friendly is regenerative grazing?', *Frontiers in Ecology and Evolution* 9 (816374): 1–9.

Naylor, Rhiannon, Alice Hamilton-Webb, Ruth Little and Damian Maye. 2016. 'The "good farmer": Farmer identities and the control of exotic livestock disease in England', *Sociologia Ruralis* 58 (1): 3–19.

Newton, Peter, Nicole Civita, Lee Frankel-Goldwater, Katharine Bartel and Colleen Johns. 2020. 'What is regenerative agriculture? A review of scholar and practitioner definitions based on processes and outcomes', *Frontiers in Sustainable Food Systems* 4 (577723): 1–11.

Overstreet, Katy K. 2018. '"A Well-Cared-For Cow Produces More Milk": The biotechnics of (dis)assembling cow bodies in Wisconsin dairy worlds', doctoral thesis, University of California, Santa Cruz. Accessed 28 February 2022. https://escholarship.org/uc/item/3q05z3wb.

Pasture for Life. 2021. *Certification Standards for Ruminant Livestock and Products from Ruminant Livestock*. Accessed 10 May 2023. https://www.pastureforlife.org/media/2022/03/PfL-Standards-Version-4.3-Feb-2022.pdf.

Rebanks, James. 2020a. 'How to save British farming (and the countryside)'. Accessed 23 May 2022. https://unherd.com/2020/09/how-to-save-british-farming-and-the-countryside/.

Rebanks, James. 2020b. *English Pastoral: An inheritance*. London: Penguin.

Smith, Jo, Bruce D. Pearce and Martin S. Wolfe. 2013. 'Reconciling productivity with protection of the environment: Is temperate agroforestry the answer?', *Renewable Agriculture and Food Systems* 28 (1): 80–92.

Soil Association Scotland. n.d. 'Higher output with regenerative grazing'. Accessed 1 June 2022. https://www.soilassociation.org/our-work-in-scotland/scotland-farming-programmes/resources-for-farmers/grassland-management/higher-output-with-regenerative-grazing/.

SRUC. n.d. 'Regenerative agriculture: Integrating livestock, farming for a better climate'. Accessed 1 June 2022. https://www.farmingforabetterclimate.org/soil-regenerative-agriculture-group/regenerative-agriculture-integrating-livestock/.

Sutherland, Lee-Ann and Ika Darnhofer. 2012. 'Of organic farmers and "good farmers": Changing habitus in rural England', *Journal of Rural Studies* 28 (3): 232–40.

Teague, Richard and Urs Kreuter. 2020. 'Managing grazing to restore soil health, ecosystem function, and ecosystem services', *Frontiers in Sustainable Food Systems* 4 (534187): 1–13.

Thomas, Virginia and Angela Cassidy. 2022. 'Practicing engaged research through pandemic times: Do not feed the animals?', *Journal of Science Communication* 21 (2): 1–20.

Vigors, Belinda, Francoise Wemelsfelder and Alistair B. Lawrence. 2023. 'What symbolises a "good farmer" when it comes to farm animal welfare?', *Journal of Rural Studies* 98: 159–70.

Werth, Samantha. 2020. 'The biogenic carbon cycle and cattle'. Accessed 27 March 2023. https://clear.ucdavis.edu/explainers/biogenic-carbon-cycle-and-cattle.

Wheeler, Rebecca, Carol Morris, Matt Lobley and Michael Winter. 2018. '"The good guys are doing it anyway": The accommodation of environmental concern among English and Welsh farmers', *Environment and Planning E: Nature and Space* 1 (4): 664–87.

White, Courtney. 2020. 'Why regenerative agriculture?', *American Journal of Economics and Sociology* 79 (3): 799–812.

Wildwood Trust Devon. 2023. 'Wildwood shop Devon experiences, Wildwood Trust'. Accessed 11 May 2023. https://shop.wildwoodtrust.org/devon-experiences/.

Wilkie, Rhoda M. 2005. 'Sentient commodities and productive paradoxes: The ambiguous nature of human–livestock relations in Northeast Scotland', *Journal of Rural Studies* 21 (2): 213–30.

Wilkie, Rhoda M. 2010. *Livestock/Deadstock: Working with farm animals from birth to slaughter*. Philadelphia: Temple University Press.

Wilson, Geoff A. 2001. 'From productivism to post-productivism … and back again? Exploring the (un)changed natural and mental landscapes of European agriculture', *Transactions of the Institute of British Geographers* 26 (1): 77–102.

Wilson, Geoffrey Alan. 2007. *Multifunctional Agriculture: A transition theory perspective*. Wallingford: CABI.

8
The adventures of a birch branch; or, a narrative ethnography of browse feeding at the Highland Wildlife Park

Alexander Mullan

Introduction

This chapter seeks to understand the kinds of relationships that are created through human–animal feeding, what resource networks are used to sustain them, and how they bind people, animals, and environment together. Answers to these questions are explored through an ethnographic account of fieldwork that took place during the summer of 2022 at the Highland Wildlife Park (HWP), part of the Royal Zoological Society of Scotland (RZSS), located west of the Cairngorms in the Spey valley. The writing is informed by conversations between the author and the park's animal keepers, as well as members of the local community, in particular 'Ken', who regularly supplied the park with browse (vegetation such as branches or leaves that are given as animal feed). Informants have been anonymised as far as possible, given the naming of their workplace.

The chapter is written in the style of an early modern 'it-narrative', a form of writing developed by eighteenth-century novelists to explore a globalising, interconnected world through the 'eyes' of an inanimate object or animal that traversed it. The title of this chapter is inspired by Charles Johnstone's *Chrysal; or the adventures of a guinea* (Johnstone 1760). It-narratives have been described as a way of uncovering 'how subjects and objects animate one another': by giving the object some form of agency – often as the main character – these stories provide a perspective that is otherwise inaccessible in other literary forms (Brown 2003). 'Accessing an inaccessible perspective' is remarkably reminiscent of Ogden, Hall, and Tanita's understanding of multispecies ethnography as a way of peeking through the cracks or clarifying blurred boundaries between human and animal (Ogden et al. 2013). This similarity inspired the use of an archaic literary form to explore the interactions between human, animal, and plant actors, thereby gaining a new perspective on this feeding interaction. Just as Anna Tsing used matsutake mushrooms to explore the world of global capitalism, this chapter uses birch to investigate the local economies and ecologies of localised human–animal feeding practices in captivity (Tsing 2015). By following the journey of a plant, this narrative opens up a new way of interpreting and understanding human–animal relationships as situated in local resource networks.

The idea behind this way of framing the research is to encourage and explore new ways of dealing with the difficulties that can arise from studying human–animal interactions. It is not seamless, however, and requires a slight suspension of disbelief on the part of the reader. The birch branch telling the tale has to have some knowledge of academic citation practices in order to produce a legitimate chapter for this volume. Shared themes between this chapter and others in the volume are present throughout, but not as elegantly signposted as they might be in a typical authorial voice. Thomas's discussion of anthropogenic and captive feeding (Chapter 11), or the context of Cooper and Kitchener's work on morphological changes in zoo animal populations (Chapter 9), both relate to the themes present here, but are not easily referenced by a stick from Scotland. However, the aim of this chapter is to illuminate the experience of the food, and highlight the utility of taking alternative perspectives in human–animal research.

The adventures of a birch branch

I am a birch branch. It is not often that my kin or I are the focus of a story like this, despite our frequent presence. This is the tale of a journey I took – from a home I did not expect to leave – a short distance across a valley to a new place both exotic and familiar. While there was much I did not recognise, there was also plenty I did, for my new home is deeply tied to the landscape around it.

I was taken to become food for the animals who also live there. In some ways, of course, I have always been food, for the bugs and the mammals who roam freely through the same valley. But this change was significant. Previously I had been food as part of an ecosystem; now I was becoming food as part of an economic system. My identity changed thoroughly. I was object-ified. But before we treat with that,

I will take you back to a stand of trees on a small ridgeline overlooking the Insh marshes.

The Insh marshes

My home was on the side of a hill: a line of trees, mainly birch, planted in a sheep field along an old drystone wall. The hill rolled away beneath me down to the valley floor – the Insh marshes, now a nature reserve. Across the valley from my stand was the Highland Wildlife Park. The park was mostly hidden by other trees and the gentle slopes of the land, but the residents on this side could sometimes hear voices on the tannoy drifting across at closing time. Back over the drystone wall was a lane, separating my field from another, and a short way back up the lane was a village. The village was small, a few houses strung out along the road and others behind them down narrow tracks. Tree coverage was patchy at this level, solitary sentinels and sparse copses; there was denser woodland up the hill behind the houses. There were gardens too, of course. Although these were working farms and smallholdings, there were also cottages and houses for local residents, as well as holiday lets for those coming to spend a few days in prime hiking country.

The village had a friendly and communal spirit, which had always been evident to me. The field in which I stood was owned and tended by two different people: Ken, who did the manual work, helped out the owner who was no longer able to do the heavy lifting or take care of the daily tasks that the place required. Ken also cut logs for one neighbour and cleared brash for another, felled and pruned trees for local landowners, and happily lent a hand to whoever needed it. Some of the brash that he cut ended up getting donated to the park – and he was by no means the only one who did this. The park, by way of the keepers, had built up a network of regular donors, who would often share contact details with friends of theirs to spread the word organically – garden waste is animal food! My village in particular had several contributors, gardener couples as well as smallholders, who donated their cuttings. I could see one couple's garden from my vantage point, and I had observed their interactions with keepers from the park a week or so before my own turn came.

The day in question was in late August; a large flatbed van threaded the narrow gap between two stone gate posts and swung around into the gravel driveway in front of the cottage. There was already a large pile of branches on the lawn – whippy bits of willow and some thicker boughs of birch. The keepers set about their work immediately, hefting the branches up into the bed of the truck. Eventually the couple saw them from the window and appeared in the driveway to lend a hand. She was dressed for indoors, while he wore an old pair of gardening overalls. The pile was shifted in a matter of moments, and then there was time for a chat. 'Which animals will get it', the couple enquired, 'and how long will it last?' 'Gone within a day', was the keeper's reply. The wife looked amazed, but the husband seemed almost downhearted: 'Trivial', he said, before the keeper, whose name was Lauren, could reassure him that without donations like this they'd never get anywhere. She had a second person with her – a researcher, I would later learn – who watched on with interest as the conversation unfolded and made some brief notes when it was done. This was one of the couple's first donations and they seemed keen to continue, clearly finding some value in the relationship. They pointed out some other trees which they intended to work on and offered to give a call when they were ready. The *leylandii* was rejected, although there was some chat about whether a different team could use it as cover for the wildcats. The alder, however, was readily accepted. 'Thicker pieces are more useful', the keepers said, implying that if he was just going to be giving the tree a quick trim, it wouldn't be worth their while driving out for it. They said their thank-yous and left, heading back to the park to deliver their payload. I got the feeling the alder might be ending up in garden waste.

  This couple had been put in touch with the park by Ken, who was clearly keen to be active in his small community. He was originally from the north of England, but had been coming here for thirty years and knew it well. In the early days, he just used to ring up the park and offer to donate what he had cut down, but now he had the inside track with the keepers and could text them as and when he had anything they might like. He got the idea from another local man who would pass through the area with a barrow of rabbits: he offered them to both the park and the locals, by way of an honesty box system, and had obviously told Ken that the keepers were always on the lookout for things they could feed out to the animals. In those earlier days, Ken used to get the odd little perk as well, like a quick tour of the reserve while the rest of the park was emptying. That didn't happen these days, but the keepers were good about sending him photos of the animals eating whatever he had just donated. 'It's nice to see you're doing some good', he said.

  Whether this would happen to me or not, I was not sure, but the threat was becoming more pressing. Ken had been down the lane a lot more often in recent months. He was converting a building just next to the track, which he wanted to turn into a holiday cottage. It was a nice wooden structure, cosy and just isolated enough that holidaymakers escaping for a week of walking would feel relaxed and surrounded by nature. What it lacked, however, was a view. I know this because I had the view it wanted, which is ultimately what led to my demise.

The felling was a quick and methodical process, if noisy and violent. The sheep made themselves scarce as the roar of the chainsaw echoed across the valley. Ken used his chainsaw to carve branches from the trunk, descended, and dragged them swiftly out of the field into the lane. He left stacks of neat logs to one side, and lengths of gnarled branches, covered in thick beards of moss and lichen, in an open space by the wall. Within a short space of time his work was complete, a short stump and a carpet of sawdust the only untidy details he did not address. It was some days later that I learned more about the reasoning behind the processes. The researcher that I had seen in the couple's garden only the week before returned to speak to Ken. He was dressed in park uniform, like one of the keepers, but he was interested in what the animals ate and where their food came from and had come to discuss the browse that got donated to them. In fact, he would return again a few days later with a keeper to load me into that very same van I had seen on his first visit. From their conversation, held in the lane looking out over the marshes, I began to learn.

Ken always checks the browse he cuts in case it contains any other vegetation that would be bad for the animals' health. Lilac, he said, is a regular offender. He is watchful of cherry as well: although no one from the park has ever told him that it is off limits, 'it makes [his] sheep queer' and he is wary of it. Cyanide in the bark perhaps, he suggests, same as apple pips. Fine if you're eating only a few, but these 'grazers' get through such an amount that it could do them some harm. He is also conscious of the health risks that the browse might pose by being in fields that hold livestock. Similar to the situation I found myself in, there are some trees he has his eye on that are in fields grazed by sheep, which carry a risk of orf. This was not a term the researcher was familiar with, and we were both grateful for Ken's explanation: orf is a virus that causes lesions and pustules in both animals and humans, and can be caught not only from infected animals but by contact with contaminated soil or vegetation. Sheep show it round their lips, whereas humans normally get it on their hands. His wife had had it. He said that if he does cut things down in fields where there is a risk then he drags the pieces out immediately, to prevent them from being contaminated by prolonged contact with the ground. He was aware that some animals in the park currently had orf and that it was a worry for the keepers – not only in terms of how they would treat it, but also to know how it got in, and what they could do to prevent it in future.

They moved on to discuss the 'why' of it all. It was fascinating to me, though galling, that I was cut down for a view. Ken explained to the researcher that there were two real reasons my tree had had to go: firstly, I obstructed the view from the main room of what was to be his holiday let, and secondly, if the frequency of storms in the area continued to rise there was a risk of damage from my tree to either his property or a neighbour's. The storms had been vicious, and there was clear evidence of this all around. Across the valley in the Highland Wildlife Park, a huge swathe of trees had come down causing damage to structures and blocking access for some of the animals on the reserves. Urgent repairs had been made, but most of the trees lay untouched for some time – presumably for lack of available labour. This potential risk does support Ken's position, but the real spark was his need to provide for future guests. The local tourist economy of the area is really what led to my felling. Ironically, too, because local tourism is also what ultimately drives the Park's need for browse: without visitors the Park would not be able to keep its animals for long. That pool of visitors is the foundation that allows it to focus on welfare and conservation – the animals rely on tourists just as locals do.

Once the operation was finished we were all stacked in a pile neatly at the side of the grassy lane that ran down between the fields, next to the wood-clad holiday cottage that Ken was in the process of renovating. It was now time to be loaded up for transport to the park.

Labour and logistics

The transport laid on for us was a flatbed truck. It was a maintenance vehicle that the keepers were allowed to borrow for big jobs like this, when their usual pick-ups would have demanded multiple trips to get the job done. They reversed it down the narrow lane and pulled up next to my pile. Lauren, the keeper, and the researcher who spoke to Ken stepped out and began loading us up, tossing branches in one way and then the other, alternating which end the leaves were at to try to keep a level load. I was amazed at how much went in: the more that goes on top, the more compressed those at the base become, until the level reaches the height of the cab. In this case, that was all that was needed, but the keeper explained that, if necessary, they could tie this load down with ropes and add a little more on top, before securing the whole thing with yet more ropes. It takes a good deal of work, levering the ropes and rocking the van, to get them pulled down tight enough that everything will stay in place on the drive back. Lauren explained why it was important to get everything loaded properly, not just for an easier drive, or to comply with the law, but to be a good citizen and protect other drivers or pedestrians from stray branches. With all the leafy ends tucked in they were ready to return. Not much was left on the side of the lane: a few leaves that had been ripped free, some flecks of moss and lichen that had been unmoored from the bark, and some sawdust and chippings from the cutting process. Nothing that would not blend in or blow away after a few days.

All of this work is a trade-off, I heard Lauren explain. There are two alternatives: one is reasonable, the other is not. The keepers can harvest browse on site, from the various plantations and stands of trees that grow in the reserve and around the service road of the park. This strategy is perfectly viable, and the keepers regularly spend their afternoons harvesting this stock, but it depletes what is essentially a reliable back-up resource if other wellsprings of browse dry up. The second strategy would be to buy it in from commercial sources, as they do frequently in Edinburgh for the zoo. To buy in everything, however, would be ruinously costly, not to mention nonsensical given this ready local supply. Sometimes there may be no alternative, in winter for example, or when a particularly specialist fodder is required, but this is rare.

As Lauren told the couple in their garden on that first occasion, most browse that the park feeds out to the animals is harvested and distributed on the same day. Thus, almost as soon as we had arrived in the yard back at the park, we were off again into the main reserve. We had paused only briefly for Lauren to check with her team leader about where exactly we should be taken. Reindeer and elk[1] was the decision they arrived at, as these two were top priority species for browse when there is less to go round.

The animals

As we drove away from the yard and into the park, I got a sense of the layout of the place. It reminded me a little of ripples stretching out over a pond: at the centre is the visitor car park, ringed by buildings like the café and visitor centre; around these are walk-round enclosures and the

---

[1] Elk (*Alces alces*), known in North America as moose, not the North American elk *Cervus canadensis*.

stores yard; stretching farther up the hill to the back is the main reserve, in which the red deer, bison, wild horses, and elk live; still further are the wildcats, and the wilder cats – Amur leopards once destined for release into the wild. Around all of this is a perimeter fence, demarcating hillside and farmland, if not entirely separating it. It was to the main reserve that we went, driving against the flow of traffic to the back entrance of 'Wolf Wood', where the reindeer lived. For members of the public, reindeer viewing takes place on a long, wooden gangway, which kinks out at the end to give a view over the reserve. The keepers work in a staging area of sorts, off the end of the walkway, which also separates the reindeer show forest from their more private area to the rear. It was well into the rut at this time of the year, so Sven, the male, was split off from the rest of his family until such time as he'd calmed down. He became a different animal in the rut, no more a gentle giant, but a heavyweight combatant ready to take the fight to anyone close to him. Lauren was explaining that, until their recent transfer to another organisation, his sons had been housed off-show just below him, and he had been pacing the back fence as if looking for a way to get down to them. She did not like their chances at that reunion. Feeding Sven in this state was not so easy, given the impossibility of going into the forest with him. On this occasion though, he still had browse left over from the day before, so the problem could be kicked over to tomorrow. We went instead to mother and daughter, who were nearby.

    They were currently being housed in part of the camel yard: a sandy paddock accessible from the rear and side of the camel house, fenced off from the main reserve. The van slowed so that Lauren could check to make sure that there were no horses or bison close to the gate that they needed to drive into. Once they were reassured, the researcher hopped out and opened the gate, still keeping a wary eye out. Although not exactly a delicate operation, there were a few seconds when the van was reversing into the gap during which interested horses, seeing fresh leaves and a chance to get in somewhere they shouldn't be, might make a rush for the browse pile. Once the flatbed and both people were fully inside it was time to unload. A second gate stands almost exactly a van's length behind the first, and it was from this cramped 'porch' that the food was to be delivered. Both the reindeer had come over expectantly, curious about what they were about to receive.

    Lauren decided to use the thinner stuff here and take the thicker branches for the elk. Having undone the ropes, they untangled the smaller branches from the pile and began tossing them over the gate into the paddock. They take care not to hit the reindeer, especially the younger one who is only a year or so old and still quite small. The two deer are often drawn to the most recently thrown branch – a grass-is-always-greener mentality – so a sort of game develops where the reindeer are led one way by the promise of fresh lichen, thereby opening up a space on the other side in which the keepers have a clear shot to throw the next branch. Sometimes mother and daughter split up, making it slightly harder to aim for the gaps.

I was destined for the elk, so stayed on board the van, giving me a good view of what happened next. The keepers entered the yard while the reindeer were distracted and began to drag the branches towards some of the older piles that were scattered around the space. The idea was that the reindeer, and before them the camels and horses, have a handful of feeding stations which they can rotate through. Older browse lies at the bottom, turned grey by the elements and stripped of bark and leaves over many return visits. The newer branches were placed on top, with as many of the leaves facing upwards as possible, to make them more easily accessible. The keepers also turned over some of yesterday's browse, where the leaves may have been trampled into the ground or lost in the tangle of older sticks. At some point, the reindeer followed their desired piece away from the gates towards the larger piles, and flitted between them seeking out the best leaves and bits of lichen. And there was lichen aplenty: the trees that were felled to give future holidaymakers their view had stood for a long time in clean, damp air above the marshes.

Once everything was moved, we set off again around the main reserve to get to the elk. Named Ash and Raven, they were still young and somewhat reclusive. Partly they were getting to know each other, and as relatively new residents of the park, they were getting to know their environment as well. They were the farthest from the car park in a separate area of the reserve that was fenced off and guarded by cattle grids, which prevent other animals on the reserve from getting in, and the elk from getting out. At the rear of their enclosure, back from the road, there are some mounds they can lie behind, obscured in the long grass. From the cab, I heard Lauren commenting that they were often the hardest animals to see in the reserve, which bothers visitors on occasion. The alternative, giving them nowhere to hide, is not really conscionable in a modern zoo (Willis 1999). The elk had two main feeding areas, one very close to the entrance, where visitors' cars drive in, and the other further along the track where it bends and begins to turn back towards the main reserve. We were driving in against the flow of traffic, so came to this secondary feeding station first. The van paused, but the heads of the elk raised up and they caught a glimpse of us. The decision was made not to risk it and we drove on: two young female reindeer are no great threat to the keeper's safety, but being in the proximity of a rutting elk and his companion is not to be risked unless safely inside the van. They quickly sped on to the first feeding area, a large pile of browse on either side of the road. With a sense of urgency they got out of the cab and Lauren told the researcher to get a few of the bigger branches and put them in the 'tree feeders'. So it was that I found myself standing upright in a length of submerged drainpipe just off the side of the main pile. Another branch was shoved in as well, so that we were locked more securely in place, less likely to be torn straight out of the ground by vigorous foraging. Having once been part of a tree, with a view down over the marshes, I now spent several days simulating being a tree of my own. Not only did I improve the aesthetics of the feeding piles and add a degree of naturalism, but the elk were also able to eat with their heads up, as if foraging from low-hanging branches. My leaves were ripped off, bark stripped, and then I was ignored in favour of other opportunities.

Similarly to what happens with the reindeer, when new browse is brought in it is simply placed on top of the barer patches of the old stuff. Yesterday's still-leafy patches were turned right-way-up, and that was that. After a day or two I was taken down and replaced in the tree feeder by fresher branches with their greenery intact, my sparse twigs added to the upper layer of the piles.

Eventually, after weeks or months, branches reach the deeper insides of the pile and are inaccessible to either keeper or animal. One day, without much warning, a yellow forklift came and hoisted our pile up into the air. We set off again, back towards the camel yard where the reindeer had been all that time ago. We stopped just outside, next to one huge pile that stood at the side of the road where the ground fell away into the centre of the reserve. This pile was much older: most was grey and brown, with hardly any leaves left on them. The forklift added its load to the pile and drove on.

Life after feeding

This is where my story shifts again, returning to something more like the ecosystem that I was part of on that ridgeline overlooking the Insh marshes. The piles take on new roles in the main reserve, with so many different animals roving around, and new social interactions to contend with. We were largely left to our own devices, with little interference from members of staff. Visitors paid us little mind, far more interested in the animals that stand around or graze from the grass nearby. From my earlier vantage in the elk enclosure I had seen these piles serving two distinct purposes: one of feeding, the other social. The bison, perhaps hardier than some of the other animals at the park, would often return to browse piles that seemed long dead. It may be that the flavour of the bark had changed in that time, or the mineral content, and they gained some extra nutritional dimension over time. Perhaps new mosses and slimes began to grow on the bark while it languished beneath the sun and rain. There is a circularity to this: one type of food creates more food. As it decomposes, or is eaten and returned to the earth as bison waste, it replenishes the nutrients in the ground, helping the grass to grow, which in turn may be grazed by bison or deer and enter once more into the cycle.

The other benefit is behavioural: in the rut, when the red deer stags become more aggressive and territorial, those old branches provide an outlet for them to thrash their antlers into, and a barrier for females and younger males, as well as other animals, to hide behind. The behavioural benefits of browse-eating are well attested, but mainly through the process of consumption (Baxter and Plowman 2001; Kawata 2008). Browse as a socio-behavioural catalyst is not well-studied, especially not in mixed-species habitats (Plowman 2013). Once when they were doing their rounds, I heard a different keeper tell the researcher about a time when a senior zoo official told them to remove all of the old piles. In the rut that year, they lost two red deer hinds to antler stab wounds, one male was attacked, and one of the female Przewalsksi's horses was gored down her flank. It had never been like this before, and the keepers put the change entirely down to the lack of dead branches that the stags could thrash their antlers on. Since then, of course, the piles have returned, and they are there to stay for the foreseeable future. In the past, most of the piles were burnt in place once they got too old or unmanageably large. A new one would be started on some clear ground nearby.

That has been my journey across a few miles of the Cairngorms. I have seen more than most and much more than I ever expected. These opportunities were afforded to me completely by chance – that the man who looked after the field I stood in wanted his future guests to have a better view. The park's policy of animal care meant that I would likely end up with one of the priority species, but it was their desire for some naturalistic presentation and my own thickness of branch that meant I was sent to the elk rather than the reindeer. Aesthetic preference had been put aside in the decision to keep the browse piles in situ, rather than removing them, and this afforded me a new life with the bison and the horses on the opposite side of the reserve.

A memoir

Although my journey has been physically short, conceptually I have come full circle. As I rot here, the goodness that remains in me will feed the earth, grow the grass, and begin the cycle of nutrient transfer again. Birds that move through the reserve will carry nutrients elsewhere in the park, as will rodents, or even the keepers themselves. New plantations will grow, more branches will be harvested, and more animals will be fed – both deliberately and environmentally.

I began by describing to you the landscape in which I stood for so many years, and in many ways this has been entirely a story of landscape. About the way that the park – a local organisation that is home to some rather non-local animals – embeds itself into both the physical and social environment by its feeding practices. As well as a food item, both willing and unwilling, I consider myself to be a social engineer (Kirksey and Helmreich 2010). My journey has brought communities together, fostered relationships between local people and the park that also calls this valley home. I have brought people and animals together too – not just the people who feed them, who get to watch from up close as reindeer skip about them looking for the best bits of lichen, or elk loom out of the long grass and trot pointedly towards a fresh batch of leaves, but also those people who donate their green waste to the park. I may not be the star of the photos the keepers send back to the donors – it tends to be the animals that take centre stage – but I know, and now you do too, that it was me that staged the whole event.

Finally, I would like to reflect on this text I have written, as I do not think there are many like it. Mine is not a viewpoint that is often heard, so I hope my tale has given you cause to reflect on these practices from a different perspective. If they are told at all, these tales will often privilege the human perspective, rather than focusing so intently on the animals and non-human actors who play their parts with equal vigour. Finding a way to tell it in a new way is a struggle, as many have pointed out before (Fudge 2000), so I am glad to have found this opportunity to make my voice heard. Especially where feeding is concerned, one loses so much of the process, the deep relationships and the networks that become embedded in the landscape. Such knowledge is largely inaccessible without the kind of observations I have been able to make. This is a sustainable and environmental feeding practice that serves a purpose beyond just nutrition: it brings together people, animals, and landscape in an act of feeding that is more complex than it may appear.

References

Baxter, Emma and Amy Plowman. 2001. 'The effect of increasing dietary fibre on feeding, rumination and oral stereotypies in captive giraffes (*Giraffa camelopardalis*)', *Animal Welfare* 10: 281–90.
Brown, Bill. 2003. *A Sense of Things: The object matter of American literature*. Chicago: University of Chicago Press.
Fudge, Erica. 2000. 'Introduction to special issue: Reading animals', *Worldviews: Environment, Culture, Religion* 4 (2): 101–13.
Johnstone, Charles. 1760. *Chrysal; or the adventures of a guinea*. London: T. Becket.
Kawata, Ken. 2008. 'Zoo animal feeding: A natural history viewpoint', *Der Zoologische Garten* 78 (1): 17–42.
Kirksey, S. Eben and Stefan Helmreich. 2010. 'The emergence of multispecies ethnography', *Cultural Anthropology* 25 (4): 545–76.
Ogden, Laura A., Billy Hall and Kimiko Tanita. 2013. 'Animals, plants, people, and things: A review of multispecies ethnography', *Environment and Society* 4 (1): 5–24.
Plowman, Amy. 2013. 'Diet review and change for monkeys at Paignton Zoo Environmental Park', *Journal of Zoo and Aquarium Research* 1 (2): 5.
Tsing, Anna Lowenhaupt. 2015. *The Mushroom at the End of the World: On the possibility of life in capitalist ruins*. Princeton: Princeton University Press.
Willis, Susan. 1999. 'Looking at the zoo', *South Atlantic Quarterly* 98 (4): 669–87.

9
You are what you eat: dietary drivers of morphological change

David Cooper and Andrew C. Kitchener

Introduction

Although animals in captivity have been, and continue to be, a source of public entertainment (for example, the flea circuses discussed in Chapter 3), modern zoos are also important centres of education, research and conservation, and connect society with the wider natural world (Tribe and Booth 2003; Roe et al. 2014). Animals within zoos are ambassadors through education and research, and potential saviours through conservation of wild populations (Chiszar et al. 1990). Yet through their captivity zoo animals are wholly dependent on human care for their health and wellbeing. Alongside shelter and water, food is at the heart of the basic needs of animals in zoos, and the diets provided to zoo animals can have lasting impacts on individuals' development, growth, reproduction, health and longevity. One of the most important impacts on captive animals is through the influence of diet on morphology, or the size and shape of the body, such as the skull and associated muscles from birth to adulthood. Understanding morphological differences between captive and wild populations is important because morphology and associated behaviours dictate function (Drake and Klingenberg 2010; West and King 2018; Morales-García et al. 2021). Differentiation in morphology is likely to lead to differentiation in fitness in the wild environment, and any maladaptations caused by captivity may hinder the success of conservation reintroduction programmes (Mitchell et al. 2021; Chirchir et al. 2022; Siciliano-Martina et al. 2022). Therefore, in addition to potential captive welfare concerns, understanding morphological variation is important under the One Plan approach to species conservation developed by the International Union for the Conservation of Nature (IUCN), which calls for the integration of all members of a species – both wild and captive – into conservation planning and management (Mallinson 2003; IUCN – SSC Species Conservation Planning Sub-Committee 2017). Additional conservation implications of how animals are fed before and during reintroductions are discussed in Chapter 11.

The potential for genetic variation through evolutionary divergence, or the domestication of zoo animals through successive captive generations, has been raised (O'Regan and Kitchener 2005), because evolutionary processes can affect the morphology of vertebrates over decadal timescales and in just a few generations (Donihue and Lambert 2015; Alberti et al. 2017). Zoos and other captive environments may provide an opportunity for evolutionary variation from wild animals through continued separation from wild populations (vicariance), in which a limited genetic pool of captive individuals caught from the wild is subject to genetic drift (founder effects). Alternatively divergence may take place through selective breeding, whether consciously or unintentionally. It is vital to maintain the evolutionary integrity of wild animal populations in captivity and to understand how management practices for zoo populations may exacerbate or reduce genetic variation and novel evolutionary trajectories (Schulte-Hostedde and Mastromonaco 2015). In captive breeding programmes this is achieved by attempting to equalise the genetic representation of all original founders so as to maximise and retain at least 90 per cent of genetic diversity over one hundred years or more. However, in this chapter we focus on non-evolutionary morphological differentiation between captive and wild mammals. Most international captive breeding programmes have been established within the last fifty years, so that differences we see in captive populations today are likely to be plastic (as a consequence of the individual's response to its environment) rather than primarily evolutionary, given the relatively few generations that have occurred since founders were taken from the wild and breeding programmes established to equalise founder genetic contributions and minimise inbreeding. This makes captive/wild comparison studies useful for understanding phenotypically plastic responses to changes in environment, with the caveat that evolutionary change is an increasing concern for each successive captive-bred generation. Of all environmental differences between captivity and the wild, diet is held accountable as the primary driver of plasticity in morphological traits (O'Regan and Kitchener 2005).

Zoos enable the investigation of phenotypic plasticity across a wide range of taxa that are not available for research within controlled laboratory experiments. As such, comparisons between captive and wild animals can act as a semi-experimental/semi-natural bridge between laboratory-controlled experiments on a limited range of animal species (most often rats and mice) and patterns of variation that occur in the wild among large, elusive or endangered species, or extinct species that have appropriate extant proxies. For example, morphological differences between captive and wild populations have been used to understand variation between different wild populations in rhesus macaques (*Macaca mulatta*) and tigers (*Panthera tigris*), where morphological plasticity occurs along environmental gradients (Arenson et al. 2022; Cooper et al. 2022). There is scope within captive/wild comparison studies to improve our understanding of subspecific taxonomy. For example, of 223 proposed subspecies of the family *Felidae* (from 40 described species by Wozencraft 2005), 220 are in part described in morphological terms, while only 143 have genetic evidence and 141 have biogeographical evidence to support or refute their validity (Kitchener et al. 2017). Historically assigned taxa based solely upon morphology may be invalidated through a better understanding of morphological drivers. Likewise, putative subspecies of the brown bear are described based on their morphological distinctiveness, despite more recent conflicting genetic evidence (Kitchener et al. 2020), and ungulate species and subspecies are extensively described based upon cranial morphometrics (Groves and Grubb 2011), somewhat in defiance of known phenotypic plasticity (Geist 1989; Applegate 2013) and lack of genetic structure within populations (e.g. Lorenzen et al. 2008). Although well-supported subspecies rely on combined genetic, biogeographical and morphological evidence (Kitchener et al. 2017, 201), there is still a legacy of taxonomy based solely on morphology, which often has little regard for the potential role of environmental variation between populations and corresponding plastic change.

Morphological changes in zoo animals as a consequence of diet have been extensively reviewed (O'Regan and Kitchener 2005; Siciliano-Martina et al. 2021; Hanegraef and Spoor 2024). The nutritional and mechanical properties of diet are significant factors leading to morphological differentiation in skulls between captive and wild animals. In this chapter we review the ultimate causes of phenotypic plasticity of skulls and mandibles (jaws) in relation to diet and how these have been examined through controlled experiments. We then consider the proximal causes of dietary and therefore morphological differentiation between captive and wild environments, and the interactions between multiple driving forces that may occur. We focus on skeletal and specifically skull and jaw morphology because of the widespread availability of modern and archaeological/palaeontological skeletons and skulls in museums, which makes changes to these structures relevant beyond captive welfare and current conservation practices. The form of the skull is also particularly susceptible to morphological changes due to dietary variation because of its role in food acquisition and processing (Lieberman et al. 2004). Although we discuss morphological change in relation to pathology, this chapter focuses on size and shape changes resulting from 'normal' growth and development, which could also be a consequence of genetic variation.

Differences between zoos and the wild

Modern zoos aim to enrich the lives of their animals, to mimic the stimuli of their wild environments and to promote natural behaviours. However, regardless of the efforts of zoological institutions, there will likely always be differences between captive and wild environments due to limitations of space and resources, and conversely due to advances in husbandry and veterinary care. Existing studies, comparing captive and wild cranio-mandibular morphology, often pinpoint dietary difference as the driving factor behind variation, but these studies can be limited in their discussion of the differences in dietary composition and their mechanical properties (Hartstone-Rose et al. 2014; Cooper et al. 2023). This is largely because of the uncontrolled experimental condition of captive versus wild studies and the multitude of factors that can lead to varying environmental conditions. This prevents us definitively linking dietary properties with morphological variation between zoos and the wild. The diets of animals in captivity may differ from those in the wild due to the difficulties of matching wild diets to those in zoos, such as the availability, cost and prevailing species-specific knowledge (an example of these practicalities is explored in Chapter 8). By bringing the natural world closer to humans, zoos tend to be situated near human population centres, often within cities. Coupled with the popularity in zoos of large mammals, which may have large home ranges in the wild, available range space within zoo environments will be smaller than that of wild environments for many captive animals. Therefore, it is difficult to match the daily, seasonal and annual activity and movements of zoo animals to their wild counterparts. However, in the wild, home ranges and activity budgets often vary greatly depending on food abundance and food distribution, so it is difficult to know how much activity is required for a 'normal' life in captivity. Similarly, by bringing global wildlife to single locations, the climate of zoos may be very different to that of an animal's natural range. Through captive husbandry and veterinary care, the risk of harm from predation, intraspecific aggression, disease, harmful parasite burdens and accidents to animals in zoos compared with the wild is greatly reduced, and is treated reactively in ways that do not occur in the wild. Although diet may appear to be only one characteristic of the many differences between captive and wild environments, it is influenced by other environmental changes; for example, space availability and environmental control may reduce energetic demands, thereby influencing calorific needs. Diet also plays an important role in harm reduction and preventative veterinary care. Therefore, in subsequent sections we focus on the ways in which diet can directly affect morphology, and then apply this knowledge to other aspects of environmental differentiation between captivity and the wild.

The nutritional properties of food

It may come as little surprise that in addition to influencing an animal's health, nutrition can play an important role in shaping an animal's morphology. However, it is important to detail the different ways in which nutrition can influence *skeletal* morphology, so that the impacts of differences between captive/wild environments and diets can be better understood. Zoo nutrition is a consequence of the quantity and quality of the food items offered, which are influenced by cost, availability, public perceptions, animal preference and associated waste (Crissey 2005). Quality refers to the *perceived* nutritional requirements of the target species, yet due to differences between these perceptions and the reality of wild diets, or due to the other aforementioned influences on available diets, it is likely that the nutritional properties of most zoo diets differ significantly from those of wild animals. The differences in the nutritional composition of diets between captive and wild animals are likely to be greater in species that are poorly studied in the wild, species with broad omnivorous diets with varying nutritional properties, species with diets that change seasonally and through their life cycle, and species with highly specialised diets where the specialised foods are difficult to provide outside their range countries (Plowman and Cabana 2019).

The nutritional properties of captive diets have changed significantly as knowledge has improved and public perceptions of the roles of zoos have changed. Primate diets have shifted from predominantly human foods, or fruit-only diets for entertainment purposes, or based on the available knowledge at the time (see Badham et al. 2000), to diets with greater proportions of browse, vegetables and pelleted food (Plowman 2013). Despite widespread improvements, captive gorillas (*Gorilla gorilla*) are often given diets with high levels of starch and sugars, which are both associated with obesity, but these nutrients are minimal in wild diets (Less et al. 2014). Captive south-east Asian colobines (*Presbytini*) in North American and European zoos are provided with a diverse range of browse, fruit and vegetables, but there is geographical variation between zoos in nutritional composition of diets, and zoo diets predominantly have lower fibre and higher protein than recorded in wild diets (Nijboer and Dierenfeld 1996). Big cats (*Panthera* spp.) have highly specialised diets of vertebrate prey (predominantly larger mammals; Hayward and Kerley 2005; Hayward et al. 2012), yet the nutritional composition of commonly used raw meat diets can vary significantly in captivity, for example from beef-based diets (57 per cent protein, 28 per cent fat) to horse-based diets (51 per cent protein, 30 per cent fat; Vester et al. 2009). Furthermore, the dietary composition – for example proportions of fibre – can affect digestibility and the uptake of nutrients such as protein (Vester et al. 2009).

The nutritional properties of diet, which can influence skeletal and skull morphology, include calorific intake as well as the proportion of macronutrients and micronutrients. Macronutrients, consumed in large quantities in the diet, are the carbohydrates, proteins and fats, which make up the principal energy source or calorific intake of food. Micronutrients (vitamins and minerals) are found in much smaller quantities but may be required for macronutrient metabolism (Olmedilla and Granado 2000) and may regulate growth during development (Savarino et al. 2021). For example vitamin D, calcium and phosphorus are essential for skeletal growth and the production of milk. Experimental data outlining the effects of nutrition upon skeletal development, health and morphology are detailed in Table 9.1.

The impacts of calorific restriction and deficiencies in macro- and micronutrients are particularly important during growth and development, and can have lasting impacts on skeletal morphology into adulthood, including pathological changes (Miller et al. 1983; Eberle et al. 1999; Searcy et al. 2004; Silva et al. 2019). It is evident that the effects of deficiencies (or excesses) of different dietary components may influence bone growth and morphology in similar ways: changes to calorific content and macro-/micronutrient composition can impact bone mass, the thickness and quality of cortical (compact outer) and

Table 9.1 Effects of nutrition upon skeletal development, health and morphology. Studies under experimental design have shown the effects of changing nutritional properties on skeletal health and morphology. Calorific quantity and proportions and quantities of macro- and micro-nutrients can have lasting impacts on skeletal shape and health that persist in adulthood. The nutritional properties of food are particularly influential during development; however they can also impact skeletal health and composition in ageing animals.

Nutritional property	Effect on morphology	Species	Reference
Calorific and protein restriction during development	Reduced skeletal size, skull and brain size, mandible length, molar diameter.	mouse (*Mus musculus*) rat (*Rattus norvegicus*) song sparrow (*Melospiza melodia*) pig (*Sus domesticus*)	(Barbeito-Andrés et al. 2016; Devlin et al. 2010; Luke et al. 1979; Mehta et al. 2002; Searcy et al. 2004)
High fat diet	Increase in bone mass.	mouse (*Mus musculus*)	(Silva et al. 2019)
High sugar & high fat diet	Reduction in bone mass and vertebral cross-sectional area.	rat (*Rattus norvegicus*)	(Zernicke et al. 1995)
Vitamin D deficiency	Decreased bone mineralisation and long bone growth.	rat (*Rattus norvegicus*)	(Halloran and De Luca 1981; Miller et al. 1983)
Calcium deficiency	Reduced skeletal bone growth, bone mass and fibrous replacement of bone.	rat (*Rattus norvegicus*) pig (*Sus domesticus*)	(Doige et al. 1975; Fujita et al. 2016)
Excess calcium to phosphorus ratio	Overgrowth of epiphyseal plates.	pig (*Sus domesticus*)	(Doige et al. 1975)
Magnesium deficiency	Reduction in bone growth and trabecular volume, and bone loss.	mouse (*Mus musculus*) rat (*Rattus norvegicus*)	(Rude et al. 2003, 1998)
Zinc deficiency	Reduced body weight and bone mass, and deterioration of trabecular bone.	rat (*Rattus norvegicus*)	(Eberle et al. 1999)
Iron deficiency	Reduced cortical bone area, femur, tibia and medullary widths.	rat (*Rattus norvegicus*)	(Medeiros et al. 2002)

Source: compiled by the authors from the works cited.

trabecular (spongy inner) bone, as well as bone length (see Table 9.1). However, the forces acting on the skeleton during development and into adulthood also affect skeletal morphology. Therefore, it may be difficult to ascertain the nature of nutritional differences between captive and wild animals by examining differences in skeletal morphology alone. This is made more difficult because the physiologies and metabolisms of zoo animals are often unknown or poorly known in comparison with commonly utilised species in experimental studies (such as mice, rats and pigs). This can cause change in shape as well as size, as nutrition affects cartilage growth, for example at specific growth plates, as well as affecting bone surface remodelling (Luke et al. 1979). Although studies have focused upon particular structures to measure the effects of nutritional differences, it is likely that nutrition has widespread effects across the skeleton rather than localised impacts at specific sites of mechanical function.

Consistent, high-nutrient diets have been implicated in larger endocranial size and brain development in captive Mexican wolves (*Canis lupus baileyi*; Siciliano-Martina et al. 2022). Nutrition may have affected the size of Indian rhinoceroses (*Rhinoceros unicornis*) in captivity (both greater and smaller than those in the wild) under differing environmental and dietary conditions, highlighting the heterogeneous nature of captive nutrition (Groves 1982). Adult body mass and speed of development were greater in zoological and research populations of chimpanzees (*Pan troglodytes*) than sanctuary chimpanzees, which is likely due to greater calorific intake and fewer energetic demands, creating a more positive energy balance (Curry et al. 2023).

The physical properties of diet

Understanding how diet can impact morphology through different forces generated by jaw musculature acting upon the skull and mandible requires interdisciplinary thinking between the fields of engineering, material sciences and biological sciences. Bone is a plastic tissue, which responds to mechanical forces acting upon it (Currey 2003). Bone formation, regeneration and degradation are stimulated by the forces that produce mechanical stress and strain as a result of muscular contraction, impact loading and gravitational forces (Ruff et al. 2006; Hart et al. 2017). The mechanical properties of diet affect masticatory muscle force production and occlusal loading (forces acting upon the upper and lower jaws from biting and chewing), and can thereby influence skull morphology (see Table 9.2). As discussed in relation to the nutritional properties of diet, bone is particularly prone to morphological change during early development if fundamental growth processes are altered (Gonzalez et al. 2011; Chan et al. 2021).

Greater stresses in bone may lead to greater intracortical remodelling (of the compact outer bone) to replace fatigued bone, thereby leading to increased bone dimensions and/or thickness of the cortex (Bouvier and Hylander 1981). This will affect the skull and mandible most at specific locations of high stress, which are likely to be species-specific due to muscle organisation and cranial geometry (Figure 9.1; see also Wroe 2007; Karamani et al. 2022). Differences in the growth pattern of the maxillary complex (upper jaw) can occur due to the mechanical forces of the masticatory muscles acting upon their origination and insertion areas (Yamamoto 1996). For example, the bony ridges associated with the origins and insertions of temporal muscles (coronoid process of the mandible, mastoid crest, nuchal crest, sagittal crest) are entirely dependent on muscular forces for their development (Washburn 1947). This occurs because masticatory muscles exert tension on the periosteal membrane of the cranial bones, leading to periosteal bone apposition at the inserting areas due to increasing stresses (Katsaros et al. 2002; He and Kiliaridis 2003). Although tension from muscular forces on the periosteum can facilitate bone formation, pressure upon periosteal surfaces, such as from adjacent muscle mass, can inhibit bone formation or induce bone resorption (Carpenter and Carter 2008; Chan et al. 2021). Therefore, bone is affected by the size and position of skeletal muscles and the forces applied to bone from the muscle body during muscle contraction. Weaker masticatory muscles may increase morphological variability of the skull, suggesting that masticatory muscle usage influences the development of facial morphology and is important for creating a functional structure for mastication (Ingervall and Helkimo 1978).

Although differences in skull structure between mammal species likely results in different responses to dietary mechanical stresses (Yamamoto 1996; He and Kiliaridis 2003; Wroe 2007; Siciliano-Martina et al. 2021), the experimental studies in Table 9.2 highlight the underlying mechanisms behind plastic morphological change in the skull and mandible in response to food 'hardness', which can be applied to a wider range of species than those studied in the research. In these studies, the term 'hardness' is often misused and generally reflects differences in food toughness (work of fracture) and Young's modulus (hereafter referred to as stiffness; see Hiiemae 2000). The processing and

Table 9.2 Experiments designed around the effects of manipulation of the physical properties of food on the mechanical response of the skull and mandible, to determine how these structures may vary plastically due to variation in stresses.

Experimental design	Effect on morphology	Species	Reference
Soft diet (to discourage molar mastication), and incisor clipped (to discourage incision)	Reduced extent of cartilage and reduced cartilage hypertrophy in intermaxillary suture.	rat (*Rattus norvegicus*)	(Hinton 1988)
Hard vs. soft diet	Soft diet: narrower dental arch (reduced mid-palatal suture growth/occlusal loading), narrower premaxilla and frontal bones (sites of masticatory muscle attachment), decreased mass of cranium and mandible, and decrease in length of angular process of the mandible.	rat (*Rattus norvegicus*)	(Beecher and Corruccini 1981a; Fujita et al. 2016; Katsaros et al. 2002; Moore 1965; Watt and Williams 1951; Yamamoto 1996)
Hard vs. soft diet	Soft diet: smaller interfrontal and interparietal widths, more slender zygomatic arches, and shorter, narrower coronoid process of the mandible.	ferret (*Mustela furo*)	(He and Kiliaridis 2003)
Hard vs. soft diet	Soft diet: ~10 per cent reduction in growth in the ventral and posterior portions of the face (where strains are highest).	rock hyrax (*Procavia capensis*)	(Lieberman et al. 2004)
Hard vs. soft diet	Soft diet: lower cortical bone thickness, greater porosity, shallower mandibles and decreased maxillary arch breadth.	rhesus macaque (*Macaca mulatta*)	(Beecher and Corruccini 1981b; Bouvier and Hylander 1981)
Hard vs. soft diet	Soft diet: greater incidence of malocclusion, reduced palate height and palate breadths, and reduction in cranial bone mineralisation.	squirrel monkey (*Saimiri sciureus*)	(Beecher et al. 1983; Corruccini and Beecher 1982)
Hard vs. soft diet	Soft diet: 25 per cent reduction in masseter muscle. Reduced facial length, arch narrowing, tooth crowding, and broadening and flattening of the mandibular symphysis.	Yucatan minipig (*Sus domesticus*)	(Ciochon et al. 1997)
Surgical removal of the temporalis muscle	No development of the coronoid process.	rat (*Rattus norvegicus*)	(Washburn 1947)

Source: compiled by the authors from the works cited.

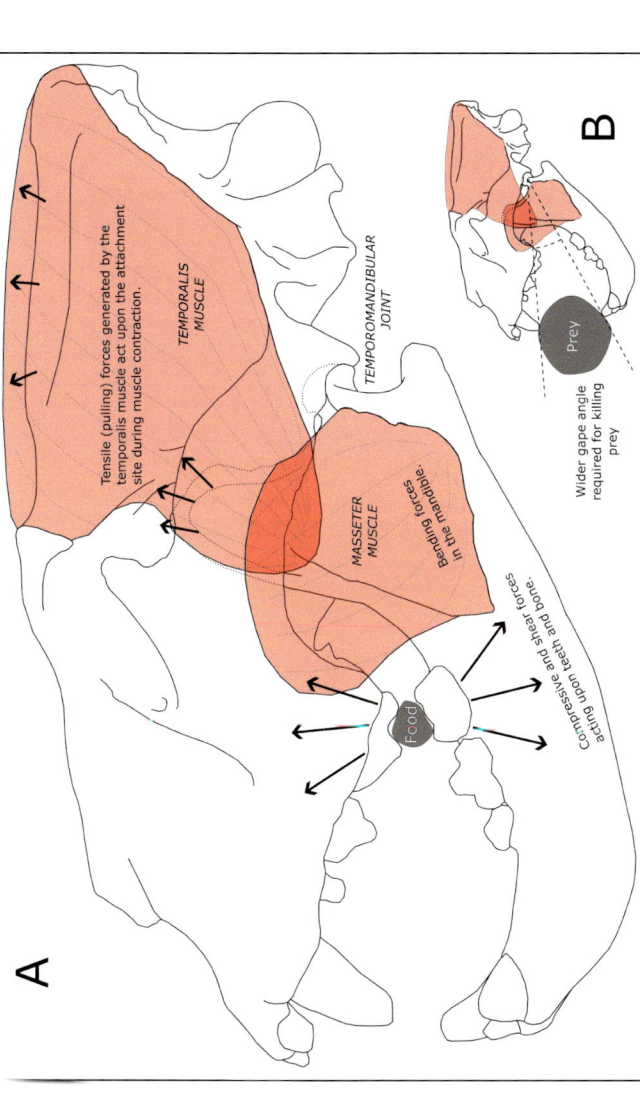

Figure 9.1 Diagram of skull and musculature. Line drawings were generated from Brutus the lion, a male that was imported from South Africa as an adult and lived at Edinburgh Zoo until he died in March 1923. Skull now in the collection of National Museums Scotland, accession number Z.1923.-02. *Source*: authors.

A. The temporalis and masseter jaw muscles generate most of the forces for chewing food. The mandible is subject to bending forces, while the carnassial teeth experience compression and shear as they slice through soft tissues. The development of the jaw muscles creates stresses in the bone of the skull, which is remodelled throughout adulthood and adulthood to resist the forces, leading to, for example, the development of a bony sagittal crest from which the enlarged temporalis muscles originate.

B. Forces are experienced at the front of the jaws as the canines are used to kill prey, which requires a larger gape angle than chewing.

consumption of food influences the development of associated functional and adjacent structures. Here we have discussed this in relation to the skull and mandible, but it may also include postcranial elements such as the neck and forelimbs. Although the nutritional properties of food may affect morphology across the skeleton, the physical properties of food affect skeletal morphology within regions subjected to mechanical forces (Moore 1965).

Table 9.2 shows experiments relating to food 'hardness' (that is, toughness) and 'softness', but to understand how stresses upon the skull and mandible may differ between captive and wild environments, it is important to understand the different physical and mechanical properties of foods (Table 9.3).

Food items must be broken down to enable swallowing and to increase the surface area to volume ratio to enable efficient digestion by enzymes. Through biting and chewing, the yield stress of a food is exceeded repeatedly as it is broken into increasingly smaller pieces, thereby increasing the surface area for digestion and decreasing the volume of each piece for easier swallowing (Hiiemae 2000; Berthaume 2016). Biological materials – such as bone, leaves and grasses – have heterogeneous and anisotropic properties (meaning their mechanical

Table 9.3 The mechanical properties of food, following definitions from Berthaume 2016. Although differences in diet are often categorised as just 'hard' or 'soft', foods have mechanical properties that affect how teeth and jaws interact with them to process them for consumption.

Property of food	Definition
Energy release rate	The amount of energy required to propagate a crack.
Toughness	The amount of energy a material can absorb prior to and during fracture.
Young's modulus (sometimes described as stiffness)	The resistance to elastic deformation under force.
Yield stress	The point where the elastic response of the material changes to plastic deformation.
Maximum strength	The maximum stress experienced by an object before fracture (could be tensile, compressive, or shear strength).
Hard/Soft	A material's resistance to puncture.
Brittle/Ductile	A material's resistance to fracture after the yield stress has been reached. A brittle material will fracture immediately, whereas a ductile object will continue to deform plasticly.

Source: compiled by the authors following Berthaume 2016.

properties are variable and change depending upon orientation; Currey 2003; Teaford et al. 2006; Zhao et al. 2020). For example, the leg of a tiger's preferred prey (such as deer or pig) is a composite made up of muscle, fat, bone, hoof, skin, fur and connective tissue, each of which has its own mechanical properties of toughness, yield stress and ultimate strength, and which likely vary with orientation, for example muscle fibre direction and location. A strong, brittle food and a weaker, more ductile food may have the same fracture toughness, but the stresses placed upon the skeletal system while processing them would be very different (Ravosa et al. 2015; Nett et al. 2021). Hard, brittle foods require high peak stress to cause elastic fracture (but low toughness), but softer, more ductile foods could require the same overall work through chewing at lower peak stresses and display high toughness (Thexton et al. 1980). Feeding on such different foods may generate differing plastic responses in the bone of the skull and mandible, leading to differences in skull morphology.

The size of a food object may not change its mechanical properties (such as smaller or larger pieces of meat or mango), but smaller food items may require fewer chewing cycles to process the object before swallowing and therefore require less energy to chew and generate lower stresses over time in the teeth, skull and mandible. The forces acting upon the teeth, skull and mandible in chewing are not only reduced in magnitude by food items being smaller, but they may obviate the role of the incisors and thereby minimise stresses on the anterior maxilla and mandible (Figure 9.1).

Food preparation by humans for consumption by zoo animals can impact both the strict mechanical properties and other physical properties of food, which may in turn influence skeletal development. Many natural foods utilised by both herbivores and carnivores are anisotropic composites, which are generally tougher and stronger than isotropic food items. The processing of foods in preparation for animal feeding reduces the composite properties of food, for example removing meat from the bone, or fruits/seeds from their husks, with the removal of tough elements (tendons, skin, husk and so on) and strong brittle elements (for instance bone and woody material). Simply chopping up foods, on the other hand, may not change their mechanical properties but still reduces the stresses acting on the skull and mandible, and the work required before swallowing. Further processing from small chunks into pellets, pastes or mince may affect the mechanical properties of food by destroying internal structures (homogenisation) and it can remove any form of elastic deformation of the food, thereby minimising toughness, stiffness and strength. For example, ground-up foods are isotropic, as plant or muscle fibres are broken down and oriented at random. Conversely, processing may lead to some foods with greater strength, for example primate pellets, leading to higher stresses in the jaw's musculo-skeletal system (Hylander and Crompton 1986).

In addition to directly breaking down food, the direct provision of (partly) processed foods by humans for feeding to zoo animals reduces the extra-masticatory functions of the skull and mandible (and the rest of the body) in food processing and may therefore indirectly discourage animal behaviours to find, acquire and process food, and clean themselves after feeding. For example, big cats in a zoo do not have to find, immobilise and kill their prey or drag it to safety from competitors, while chimpanzees (*Pan troglodytes*) do not have to dehusk fruits to obtain the food within. However, well-designed enrichment – such as the tiger feeding pole which anchors a natural food item in an elevated position (Figure 9.2; Law and Kitchener 2020) – may be able

Figure 9.2 A tiger pulling meat from a feeding pole at Glasgow Zoo. *Source*: photo by Graham Law, © Rosanne Strachan Law.

to effectively replicate the behaviours and their associated stresses on the musculo-skeletal system in captivity compared with traditional methods of feeding big cats in zoos (Table 9.4). This probably results in a more normal development of the entire musculo-skeletal system, which requires the use of evolutionary adaptations associated with species-specific behaviours. Wild marmosets (*Callithrix, Cebuella, Mico* spp.) gouge bark from trees using their incisor and canine teeth, which stimulates production of exudate gum, on which they feed (Thompson et al. 2014). In this behaviour, marmosets anchor their upper incisors against the tree bark while their lower teeth indent and fracture the bark, using their whole body to help to force the jaws to close. The forces acting on the jaws during bark-gouging probably exceed those required for processing and consuming any other dietary item, so that the lack of an outlet for this behaviour in captivity may have consequences for the normal development of teeth and jaws. Appropriate enrichment within captivity, such as including freshly cut branches of exudate-producing trees, artificial gum feeders, or allowing free-ranging animals access to trees within the zoo, may replicate this feeding behaviour (Tan and Drake 2001; Regaiolli et al. 2020), and therefore the forces acting on the jaws and neck of captive marmosets may be similar to those of wild animals.

Dietary interactions

The nutritional properties of foods influence the development and maintenance of bone across the skeleton, whereas the mechanical and physical properties of foods influence bone at specific sites related to the mechanical functioning of the musculo-skeletal system of the jaws (Kiliaridis 1989). This is simple to elucidate from experimental studies that vary a single property of diet, but there are likely to be interacting factors – between the nutritional properties of food, forces acting upon the skull and mandible, and the physical environment of an animal – that are worth considering here. For example, a less calorie-dense food source requires an animal to eat more of it to gain the same energy as a more calorific food. Therefore, the animal must spend more time acquiring, processing, and chewing low-calorie food, thereby increasing the duration of stresses acting upon the skull and mandible and the energy required to consume it, even if both food sources have exactly the same mechanical properties. If the basic needs of an animal are available within a smaller or less complex physical environment, then that animal

Table 9.4 The typical behaviours associated with the hunting, killing and eating of prey by wild tigers compared with behaviours elicited by different kinds of captive diet and the tiger feeding pole.

Behaviour	Adaptations	Processed diets	Meat	Whole or part carcasses	Tiger feeding pole	Wild tiger
Tiger moves through home range or waits in ambush	Musculo-skeletal system (limbs, paws)					+
Tiger sees, hears or smells prey/food	Senses (eyes, ears, nose)	+	+	+	+	+
Tiger stalks prey	Sense (eyes), musculo-skeletal system (limbs, paws), coloration and markings				+	+
Tiger rushes at prey	Musculo-skeletal system (limbs, paws, tail)				+	+
Tiger grabs prey with forepaws and jaws	Musculo-skeletal system (limbs, paws, claws, teeth, jaws)				+	+
Tiger kills prey	Musculo-skeletal system (limbs, paws, teeth and jaws)				+^	+
Tiger may go for 'a walk' to dissipate adrenaline	Musculo-skeletal system (limbs)			+	+	+
Tiger drags prey to quiet spot	Musculo-skeletal system (limbs, paws, teeth and jaws)			+	+	+
Tiger plucks fur or feathers	Musculo-skeletal system (paws, neck, teeth and jaws)			+	+	+
Tiger opens up carcass	Musculo-skeletal system (teeth, jaws, forelimbs)			+	+	+
Tiger eats meat and manipulates carcass	Musculo-skeletal system (teeth, jaws, tongue, paws, limbs), gastro-intestinal tract	+*	+*	+	+	+

Source: compiled by the authors.

* Denotes minimal use of musculo-skeletal system.
^ Denotes less use of canines than wild tiger.

has a lower energy requirement for locomotion than one that must navigate a complex larger environment (Neaux et al. 2022). Similarly, energy demands may change due to climate, affecting the degree of thermoregulation. Having lower energy requirements will require less food to be eaten, which requires fewer chewing cycles to process (Ravosa et al. 2015; Nett et al. 2021) and therefore lower cumulative forces are experienced by the skull and mandible. However, if the same amount of a food is eaten, then individuals with lower energy demands will have a greater calorific surplus than those with higher energy demands. It is probable that in the majority of circumstances, captive animals do not need to move about as much as wild animals to obtain their basic needs due to the provision of food, which limits the need to forage over large distances, which are limited in captivity anyway. However, variation may arise in how different institutions respond to the lower energy demands of captive animals. For example, many primate diets in captivity have been changed in recent years to ones with greater proportions of low-calorie, high-fibre browse and leafy vegetables, in an effort to reduce intake of sugars, and hence obesity and prevalence of dental pathologies, while also increasing feeding time, which elicits natural behaviours (Plowman and Cabana 2019). It is likely that in addition to these benefits, captive primates that eat lower-calorie, more-fibrous diets produce greater stresses in the musculo-skeletal system of their jaws due to differences in the mechanical properties of the diet (tougher leaves compared with soft fruits) and a need to consume greater quantities of these foods in comparison with the high-sugar fruit diets typical of twentieth-century captivity.

Non-dietary factors

Although diet is likely to play a major role in causing plastic changes between captive and wild skull morphology, there are other mechanisms that could cause differentiation. Modern zoos have a duty of care, which reduces harm through regular feeding, but also by preventing predation, reducing intraspecific aggression (by separating particular individuals) and by creating a safer physical environment through well-constructed enclosures. For example, trauma from falling and attacks from conspecifics are the primary causes of pathologies in wild chimpanzees (Carter et al. 2008), but the incidences of these can be reduced through captive management. Through veterinary care, any trauma and illness that does occur is more survivable in captivity, and so populations in zoos may live longer and their skeletons may exhibit greater numbers of accumulated pathologies (Kitchener 2023). Most skeletal pathologies are easily distinguished from normal bone development, but cranium thickness, skull diameter and viscerocranium (facial bones) structure size have been shown to continue to increase from early adulthood into later life in humans (Israel 1973). Therefore, as a consequence of greater average age at death through good husbandry and veterinary care (Kitchener 2023), the cranial morphology of the skulls of captive animals may differ in comparison to that of wild animals. The social environment of some species may also lead to differences between captive and wild populations through hormonal changes (Hanegraef and Spoor 2024). For example, it has been suggested that differences in the mandibular morphology of captive and wild Japanese macaques (*Macaca fuscata*) are influenced by changes in androgenic hormones, because shape changes between captive and wild individuals mirror shape differences between the sexes (Kamaluddin et al. 2019). Less stimulating captive environments may impact skull development due to reduced brain growth (Bennett et al. 1969; Cummins et al. 1973). Brain growth can in turn impact cranial vault growth by eliciting signals that promote lengthening of growth plates and increases in bone deposition between the flat bones of the skull before they are fused (Jin et al. 2016). Although reduced cranial volume has been noted in several captive examples (O'Regan and Kitchener 2005), brain development and size in captive chimpanzees and in stripe-faced dunnarts (*Sminthopsis macroura*) was not found to be different compared to their wild counterparts, possibly due to sufficient environmental enrichment (Guay et al. 2012; Cofran 2018). We do not yet understand the complex relationships in mammals between brain and cranial development on the one hand and jaw and jaw-muscle development on the other. Is cranial volume in captive mammals smaller because they eat softer foods, or because their cognitive development is impaired by smaller, less complex captive environments?

So far dietary and non-dietary factors that may influence skull morphology through phenotypic plasticity have been examined, as well as how these factors are likely to differ between captivity and the wild. These factors may influence morphology in similar ways – for example, a reduction in overall bone development may be caused by protein or calcium deficiencies – and so it is important to understand the life history of individuals or populations in captivity if we want to identify causal relationships between diet, environment and morphology. Captive care has evolved considerably through the twentieth and twenty-first centuries, and with digital record-keeping and diet plans, with access to software for analysing nutritional content (to identify potential deficiencies), it is increasingly feasible to assess morphological variation in relation to known diets (and nutrients) and care of individuals, rather than just of populations. There is scope to assess how changes in captive care through time, or between institutions, have impacted the morphology and health of zoo animals as a consequence of inappropriate dietary variation.

Wider implications

Experimental studies of mammals tend to focus on murids (such as rats and mice), humans and non-human primates. Zoos and museums offer the opportunity to explore plastic variation in skull morphology across a wide range of species whose cranial anatomies have evolved to adapt to a wide range of foods. For example, instead of using murid experiments to elucidate human evolution and ontogeny, Lieberman et al. (2004) used the rock hyrax (*Procavia capensis*) as a more appropriate analogue, the facial structure of which has a similar orientation to that of humans. This reasoning can be applied to other species of interest, both extant and extinct, illustrated here with examples of the continental tiger (*Panthera tigris tigris*), the giant panda (*Ailuropoda melanoleuca*) and the chimpanzee (*Pan troglodytes*).

Work on the continental tiger has highlighted the utility of studies of captive versus wild individuals to assess morphological variation across wild populations (Cooper et al. 2022). Captivity has enabled the plastic component of morphological change to be separated from genetically determined variation, because of the similarity in genetic provenance between tigers in zoos and in the wild (Cooper et al. 2023). Captive Amur tigers, originating from the Russian Far East, have shorter coronoid processes of the mandible, reduced sagittal crest heights, increased skull widths and decreased cranial volumes, while overall skull size is unchanged. Experimental studies discussed here show us that, because the overall skull size does not differ, but the measurements of structures associated with or adjacent to the masticatory apparatus do differ, the variation between captive and wild tiger skulls is most likely a consequence of the physical properties of diet, and not their nutritional properties. Zoo diets of tigers may be chopped into smaller pieces, processed into mince or pellets, or have skin, bone and connective tissues removed. This processing decreases the composite nature of the diet, its overall toughness, and other mechanical properties, while feeding enrichments may not fully replicate the behaviours and stresses of subduing prey (Cooper et al. 2023). Owing to the lower energy demands of captivity compared to the wild, captive tigers require less food and so spend less time processing food items, reducing the work carried out by the craniomandibular complex. The patterns in skull variation between captive and wild Amur tigers were mirrored between wild Amur tigers and other wild continental tigers, suggesting that morphological variation in the tiger's skull across continental Asia is largely a consequence of plastic responses to local diets. The energy demands of Amur tigers in the Russian Far East are greater than those of southern Asia due to considerably lower average temperatures and larger home-range sizes. Therefore, wild Amur tigers require more food, which requires more chewing. Food carcasses consumed over several days during the long winters of Siberia are susceptible to freezing, increasing the fracture toughness and yield strengths of meat, and resulting in more work and greater stresses in the jaw musculo-skeletal system. This difference is manifested in the development of a large bony sagittal crest for the origination of larger temporalis muscles, which are lacking in other continental and most captive Amur tigers. This study highlights the relevance of captive/wild comparison studies for unpicking the influence of plastic variation from evolutionary variation in wild morphology.

The giant panda (*Ailuropoda melanoleuca*) is a bamboo specialist and an international icon of conservation much in demand by zoos. The species has suffered a significant reduction in geographical distribution in the wild, and yet has maintained high genetic diversity (Zhang et al. 2007). Seasonal variation in bamboo quality may affect the consistency of nutritional and gross calorific intake by giant pandas in their current range (Li et al. 2017), and bamboos at the higher elevations where giant pandas now reside have lower protein and amino acid concentrations than those of lower elevations (Shi et al. 2019) that the giant panda occupied until the eighteenth and nineteenth centuries (Zhu et al. 2013; Turvey et al. 2017). The skulls of fossil giant pandas from the late Pleistocene found at lower elevations are characterised by their larger size in comparison with those of extant giant pandas (Hu et al. 2023). This follows a trend in megafaunal body size reduction from the Late Pleistocene to Holocene, which has been linked to reduced quality of primary productivity (Harris and Mundel 1974; Guthrie 1982). Differences in the shapes of today's giant pandas and their Pleistocene ancestors occur at the attachment sites of the masticatory muscles, with greater development of the sagittal crest, zygomatic arches and the ramus of the mandible (Hu et al. 2023). This has led to these fossil pandas being described as a separate species *A. baconi* (Jin et al. 2007) or as a subspecies of the extant giant panda *A. melanoleuca baconi* (Hu et al. 2023). Although modern pandas have likely lost some genetic diversity compared with their Holocene and Pleistocene counterparts (Barlow et al. 2019), given the nature of the morphological difference between *A. baconi/A. m. baconi* and extant giant pandas, it is reasonable to propose that the nutritional and mechanical properties of different bamboo diets are likely to have influenced the morphological differences between these putatively different taxa, either plastically as described here, or via rapid adaptive change under strong selection pressure from resource availability (Harris and Mundel 1974). Captive giant pandas have less seasonal variation in the nutritional properties of bamboo, with greater variety of food types and with nutritional properties that may exceed those of their extant wild counterparts (Dierenfeld et al. 1995). They are also fed a soft 'panda cake' which contains processed high-energy foods. Therefore, an assessment of the differences between captive and wild giant panda skulls and mandibles could be used to understand plastic morphological variation in response to diet and help determine the taxonomic status of extant and extinct pandas. It would be expected that the calorific demands of giant pandas in captivity would be lower, and the availability of nutrient-rich foods would be higher, than for wild giant pandas in their current range.

The fields of human health, evolution and anthropology have utilised studies of diets of hard food versus soft food to better understand skull plasticity, using appropriate species analogues, such as non-human primates and, as previously mentioned, the rock hyrax (Corruccini and Beecher 1982; Lieberman et al. 2004). One potential case is understanding the morphological transition of human skulls and mandibles following dietary changes during our evolutionary history (Lieberman et al. 2004). For example, post-agricultural human populations exhibit a tall, narrow mandibular ramus, a narrower and more elongated coronoid process and a narrower mandibular notch in comparison to those of pre-agricultural populations, but with no change in size (Pokhojaev et al. 2019). Here, it is proposed that a plastic response to differences in the mechanical properties of diet has driven this morphological change. Although laboratory studies have utilised suitable non-human proxies to provide controlled experiments on hard and soft diets and, as explained in this chapter, these are important for understanding the ultimate causes of plastic variation, there is merit in investigating morphological change in the more open experimental settings of captive care.

For example, chimpanzees are our closest living relative, with similar bite-force mechanics (Wroe et al. 2010), and they have been kept through the twentieth and twenty-first centuries in a wide range of captive conditions. Chimpanzee diets in the wild are varied, consisting of fruits, leaves, nuts, seeds, vertebrates and arthropods (Stuhlträger et al. 2019), while in captivity the diet may consist of primate pellets of much higher calorific content, high-sugar soft fruits and processed human foods (Plowman and Cabana 2019). Wild daily travel distances of chimpanzees are typically 2–4 km, whereas captive chimpanzees in indoor-outdoor zoo enclosures typically move less than 2 km a day (Ross and Shender 2016) and those in less enriched environments may move even less. The combinations of a more sedentary life, alongside higher calorific and potentially softer foods, provide an opportunity to explore the impacts of multiple processes, which have been implicated in affecting the morphology of the maxillae (Hanegraef and Spoor 2024), and could affect the wider morphology of the skull and mandible. Exploring the role of phenotypic plasticity in response to diet and environment in a great ape like the chimpanzee has implications for understanding transitions between human hunter-gatherer societies and post-agricultural societies with increased sedentary lifestyles and foods with greater calorific content.

Conclusion

This chapter has highlighted the primarily diet-related mechanisms causing phenotypic plasticity of the skull and mandible between captive and wild mammals. An understanding of the ultimate causes of morphological plasticity and how these are expressed across skeletal structures allows for better interpretation of the causes of morphological change in captivity, which can be used to improve captive care and the potential success of conservation reintroductions. They can also shed light on human health in understanding the implications of increasingly sedentary lifestyles and eating highly processed foods. This chapter highlights the opportunity that captive/wild comparison studies present for understanding variation in extant and extinct wild populations, where appropriate experimental models would be impractical or impossible, and emphasises the need for continued collecting of specimens of both wild and captive species across a wide range of taxa by museums as a future research resource. The chapter further serves to demonstrate how all animal feeding – including feeding for non-utilitarian, non-product-focused purposes – has the potential to reshape animal morphology and behaviour, reinforcing that it is crucial to further consider what, how and why we feed.

References

Alberti, Marina, John Marzluff and Victoria M. Hunt. 2017. 'Urban driven phenotypic changes: Empirical observations and theoretical implications for eco-evolutionary feedback', *Philosophical Transactions of the Royal Society B: Biological Sciences* 372 (1712): 1–9.

Applegate, Roger D. 2013. 'Book Review: Groves, C., and P. Grubb. 2011. UNGULATE TAXONOMY', *Journal of Mammalogy* 94 (1): 245–6.

Arenson, Julia L., Evan A. Simons, Monya Anderson, Andrea R. Eller, Frances J. White and Stephen R. Frost. 2022. 'Comparison of captive and wild *fascicularis*-group macaques (Primates, Cercopithecidae) provides insight into cranial form changes in response to rapid environmental changes', *American Journal of Biological Anthropology* 178 (3): 417–36.

Badham, Molly, Nathalie Evans and Maureen Lawless. 2000. *Molly's Zoo: Monkey mischief at Twycross*. London: Simon and Schuster.

Barbeito-Andrés, Jimena, Paula N. Gonzalez and Benedikt Hallgrímsson. 2016. 'Prenatal development of skull and brain in a mouse model of growth restriction', *Revista Argentina de Antropología Biológica* 18 (1): 1–13.

Barlow, Axel, Gui-Lian Sheng, Xu-Long Lai, Michael Hofreiter and Johanna L. A. Paijmans. 2019. 'Once lost, twice found: Combined analysis of ancient giant panda sequences characterises extinct clade', *Journal of Biogeography* 46 (1): 251–3.

Beecher, R. M. and R. S. Corruccini. 1981a. 'Effects of dietary consistency on craniofacial and occlusal development in the rat', *Angle Orthodontist* 51 (1): 61–9.

Beecher, R. M. and R. S. Corruccini. 1981b. 'Effects of dietary consistency on maxillary arch breadth in macaques', *Journal of Dental Research* 60 (1): 68.

Beecher, R. M., R. S. Corruccini and M. Freeman. 1983. 'Craniofacial correlates of dietary consistency in a nonhuman primate', *Journal of Craniofacial Genetics and Developmental Biology* 3 (2): 193–202.

Bennett, Edward L., Mark R. Rosenzweig and Marian C. Diamond. 1969. 'Rat brain: Effects of environmental enrichment on wet and dry weights', *Science* 163 (3869): 825–6.

Berthaume, Michael A. 2016. 'Food mechanical properties and dietary ecology', *American Journal of Physical Anthropology* 159: 79–104.

Bouvier, Marianne and William L. Hylander. 1981. 'Effect of bone strain on cortical bone structure in macaques (*Macaca mulatta*)', *Journal of Morphology* 167 (1): 1–12.

Carpenter, R. Dana and Dennis R. Carter. 2008. 'The mechanobiological effects of periosteal surface loads', *Biomechanics and Modeling in Mechanobiology* 7 (3): 227–42.

Carter, Melinda L., Herman Pontzer, Richard W. Wrangham and Julian Kerbis Peterhans. 2008. 'Skeletal pathology in *Pan troglodytes schweinfurthii* in Kibale National Park, Uganda', *American Journal of Physical Anthropology* 135 (4): 389–403.

Chan, Audrey S. M., Narelle E. McGregor, Ingrid J. Poulton, Justin P. Hardee, Ellie H.-J. Cho, T. John Martin, Paul Gregorevic, Natalie A. Sims and Gordon S. Lynch. 2021. 'Bone geometry is altered by follistatin-induced muscle growth in young adult male mice', *JBMR Plus* 5 (4): 1–12.

Chirchir, Habiba, Christopher Ruff, Kristofer M. Helgen and Richard Potts. 2022. 'Effects of reduced mobility on trabecular bone density in captive big cats', *Royal Society Open Science* 9 (3, 211345): 1–14.

Chiszar, David, James B. Murphy and Warren Iliff. 1990. 'For zoos', *Psychological Record* 40 (1): 3–13.

Ciochon, R. L., R. A. Nisbett and R. S. Corruccini. 1997. 'Dietary consistency and craniofacial development related to masticatory function in minipigs', *Journal of Craniofacial Genetics and Developmental Biology* 17 (2): 96–102.

Cofran, Zachary. 2018. 'Brain size growth in wild and captive chimpanzees (*Pan troglodytes*)', *American Journal of Primatology* 80 (7, e22876): 1–8.

Cooper, David M., Nobuyuki Yamaguchi, David W. Macdonald, Olga G. Nanova, Viktor G. Yudin, Andrew J. Dugmore and Andrew C. Kitchener. 2022. 'Phenotypic plasticity determines differences between the skulls of tigers from mainland Asia', *Royal Society Open Science* 9 (11, 220697): 1–13.

Cooper, David M., Nobuyuki Yamaguchi, David W. Macdonald, Bruce D. Patterson, Galina P. Salkina, Viktor G. Yudin, Andrew J. Dugmore and Andrew C. Kitchener. 2023. 'Getting to the meat of it: The effects of a captive diet upon the skull morphology of the lion and tiger', *Animals* 13 (23, 3616): 1–18.

Corruccini, Robert S. and Robert M. Beecher. 1982. 'Occlusal variation related to soft diet in a nonhuman primate', *Science* 218 (4567): 74–6.

Crissey, S. 2005. 'The complexity of formulating diets for zoo animals: A matrix', *International Zoo Yearbook* 39 (1): 36–43.

Cummins, R. A., R. N. Walsh, O. E. Budtz-Olsen, T. Konstantinos and C. R. Horsfall. 1973. 'Environmentally-induced changes in the brains of elderly rats', *Nature* 243 (5409): 516–18.

Currey, J. D. 2003. 'The many adaptations of bone', *Journal of Biomechanics* 36 (10): 1487–95.

Curry, Bryony A., Aimee L. Drane, Rebeca Atencia, Yedra Feltrer, Glyn Howatson, Thalita Calvi, Christopher Palmer, Sophie Moittie, Steve Unwin, Joshua C. Tremblay, Meg M. Sleeper, Michael L. Lammey, Steve Cooper, Mike Stembridge and Rob Shave. 2023. 'Body mass and growth rates in captive chimpanzees (*Pan troglodytes*) cared for in African wildlife sanctuaries, zoological institutions, and research facilities', *Zoo Biology* 42 (1): 98–106.

Devlin, Maureen J., Alison M. Cloutier, Nishina A. Thomas, David A. Panus, Sutada Lotinun, Ilka Pinz, Roland Baron, Clifford J. Rosen and Mary L. Bouxsein. 2010. 'Caloric restriction leads to high marrow adiposity and low bone mass in growing mice', *Journal of Bone and Mineral Research* 25 (9): 2078–88.

Dierenfeld, Ellen S., Xianmeng Qiu, Susan A. Mainka and Wei-Xin Liu. 1995. 'Giant panda diets fed in five Chinese facilities: An assessment', *Zoo Biology* 14 (3): 211–22.

Doige, C. E., B. D. Owen and J. H. L. Mills. 1975. 'Influence of calcium and phosphorus on growth and skeletal development of growing swine', *Canadian Journal of Animal Science* 55 (1): 147–64.

Donihue, Colin M. and Max R. Lambert. 2015. 'Adaptive evolution in urban ecosystems', *AMBIO* 44 (3): 194–203.

Drake, Abby Grace and Christian Peter Klingenberg. 2010. 'Large-scale diversification of skull shape in domestic dogs: Disparity and modularity', *American Naturalist* 175 (3): 289–301.

Eberle, J., S. Schmidmayer, R. G. Erben, M. Stangassinger and H.-P. Roth. 1999. 'Skeletal effects of zinc deficiency in growing rats', *Journal of Trace Elements in Medicine and Biology* 13 (1–2): 21–6.

Fujita, Yuko, Shota Goto, Maika Ichikawa, Ayako Hamaguchi and Kenshi Maki. 2016. 'Effect of dietary calcium deficiency and altered diet hardness on the jawbone growth: A micro-CT and bone histomorphometric study in rats', *Archives of Oral Biology* 72: 200–10.

Geist, Valerius. 1989. 'Environmentally guided phenotype plasticity in mammals and some of its consequences to theoretical and applied biology'. In *Alternative Life-History Styles of Animals*, edited by Michael N. Bruton, 153–76. Dordecht: Kluwer.

Gonzalez, Paula N., Benedikt Hallgrímsson, Evelia E. Oyhenart. 2011. 'Developmental plasticity in covariance structure of the skull: Effects of prenatal stress', *Journal of Anatomy* 218 (2): 243–57.

Groves, Colin P. 1982. 'The skulls of Asian rhinoceroses: Wild and captive', *Zoo Biology* 1: 251–61.

Groves, Colin P. and Peter Grubb. 2011. *Ungulate Taxonomy*. Baltimore: Johns Hopkins University Press.

Guay, P.-J., M. Parrott and L. Selwood. 2012. 'Captive breeding does not alter brain volume in a marsupial over a few generations: Marsupial captive breeding and brain size', *Zoo Biology* 31 (1): 82–6.

Guthrie, R. Dale. 1982. 'Mammals of the mammoth steppe as palaeoenvironmental indicators'. In *Paleoecology of Beringia*, edited by David M. Hopkins, John. V. Matthews, Jr., Charles E. Schweger and Steven B. Young, 307–26. London: Elsevier.

Halloran, Bernard P. and Hector F. De Luca. 1981. 'Effect of vitamin D deficiency on skeletal development during early growth in the rat', *Archives of Biochemistry and Biophysics* 209 (1): 7–14.

Hanegraef, Hester and Fred Spoor. 2024. 'Maxillary morphology of chimpanzees: Captive versus wild environments', *Journal of Anatomy* 244 (6): 977–94.

Harris, Arthur H. and Peter Mundel. 1974. 'Size reduction in bighorn sheep (*Ovis canadensis*) at the close of the Pleistocene', *Journal of Mammalogy* 55 (3): 678–80.

Hart, N. H., S. Nimphius, T. Rantalainen, A. Ireland, A. Siafarikas and R. U. Newton. 2017. 'Mechanical basis of bone strength: Influence of bone material, bone structure and muscle action', *Journal of Musculoskeletal and Neuronal Interactions* 17 (3): 114–39.

Hartstone-Rose, Adam, Hannah Selvey, Joseph R. Villari, Madeline Atwell and Tammy Schmidt. 2014. 'The three-dimensional morphological effects of captivity', *PLOS ONE* 9 (11, e113437): 1–15.

Hayward, M. W., W. Jedrzejewski and B. Jedrzejewska. 2012. 'Prey preferences of the tiger *Panthera tigris*', *Journal of Zoology* 286: 221–31.

Hayward, Matt W. and Graham I. H. Kerley. 2005. 'Prey preferences of the lion (*Panthera leo*)', *Journal of Zoology* 267 (3): 309–22.

He, Tailun and Stavros Kiliaridis. 2003. 'Effects of masticatory muscle function on craniofacial morphology in growing ferrets (*Mustela putorius furo*)', *European Journal of Oral Sciences* 111 (6): 510–17.

Hiiemae, Karen M. 2000. 'Feeding in mammals'. In *Feeding*, edited by Kurt Schwenk, 411–48. London: Elsevier.

Hinton, Robert J. 1988. 'Response of the intermaxillary suture cartilage to alterations in masticatory function', *Anatomical Record* 220 (4): 376–87.

Hu, Haiqian, Haowen Tong, Qingfeng Shao, Guangbiao Wei, Haidong Yu, Jingsong Shi, Xunqian Wang, Can Xiong, Yu Lin, Ning Li, Zhaoying Wei, Ping Wang and Qigao Jiangzuo. 2023. 'New remains of *Ailuropoda melanoleuca baconi* from Yanjinggou, China: Throwing light on the evolution of giant pandas during the Pleistocene', *Journal of Mammalian Evolution* 30 (1): 137–54.

Hylander, W. L. and A. W. Crompton. 1986. 'Jaw movements and patterns of mandibular bone strain during mastication in the monkey *Macaca fascicularis*', *Archives of Oral Biology* 31 (12): 841–8.

Ingervall, B. and Eva Helkimo. 1978. 'Masticatory muscle force and facial morphology in man', *Archives of Oral Biology* 23 (3): 203–6.

Israel, H. 1973. 'Age factor and the pattern of change in craniofacial structures', *American Journal of Physical Anthropology* 39 (1): 111–28.

IUCN (SSC Species Conservation Planning Sub-Committee). 2017. *Guidelines for Species Conservation Planning: Version 1.0*. Gland: IUCN, International Union for Conservation of Nature.

Jin, Changzhu, Russell L. Ciochon, Wei Dong, Robert M. Hunt, Jr., Jinyi Liu, Marc Jaeger and Qizhi Zhu. 2007. 'The first skull of the earliest giant panda', *Proceedings of the National Academy of Sciences* 104 (26): 10932–7.

Jin, Sung-Won, Ki-Bum Sim and Sang-Dae Kim. 2016. 'Development and growth of the normal cranial vault : An embryologic review', *Journal of Korean Neurosurgical Society* 59 (3): 192–6.

Kamaluddin, Siti Norsyuhada, Mikiko Tanaka, Hikaru Wakamori, Takeshi Nishimura and Tsuyoshi Ito. 2019. 'Phenotypic plasticity in the mandibular morphology of Japanese macaques: Captive–wild comparison', *Royal Society Open Science* 6 (7, 181382): 1–13.

Karamani, Ioanna I., Ioannis A. Tsolakis, Miltiadis A. Makrygiannakis, Maria Georgaki and Apostolos I. Tsolakis. 2022. 'Impact of diet consistency on the mandibular morphology: A systematic review of studies on rat models', *International Journal of Environmental Research and Public Health* 19 (5, 2706): 1–22.

Katsaros, Christos, Rolf Berg and Stavros Kiliaridis. 2002. 'Influence of masticatory muscle function on transverse skull dimensions in the growing rat', *Journal of Orofacial Orthopedics / Fortschritte der Kieferorthopädie* 63 (1): 5–13.

Kiliaridis, Stavros. 1989. 'Muscle function as a determinant of mandibular growth in normal and hypocalcaemic rat', *European Journal of Orthodontics* 11 (3): 298–308.

Kitchener, Andrew C. 2023. 'The longevity legacy: the challenges of old animals in zoos'. In *Optimal Wellbeing of Ageing Wild Animals in Human Care*, edited by Sabrina Brando and Sarah Chapman, 187–225. Cham: Springer Nature.

Kitchener, Andrew C., Eva Bellemain, Xiang Ding, Alexander Kopatz, Verena E. Kutschera, Valentina Salomashkina, Manuel Ruiz-García, Tabitha Graves, Yiling Hou, Lars Werdelin and Axel Janke. 2020. 'Systematics, evolution, and genetics of bears'. In *Bears of the World*, edited by Vincenzo Penteriani and Mario Melletti, 3–20. Cambridge: Cambridge University Press.

Kitchener, Andrew C., Christine Breitenmoser-Wuersten, Eduardo Eizirik, Anthea Gentry, Lars Werdelin, Andreas Wilting, Nobuyuki Yamaguchi, Alexei V. Abramov, Per Christiansen, Carlos A. Driscoll, Will Duckworth, Warren Johnson, Shu-Jin Luo, Erik Meijaard, Paul O'Donoghue, Jim Sanderson, Kevin Seymour, Michael W. Bruford, Colin Groves, Michael Hoffmann, Kristin Nowell, Zena Timmons and Shanan S. Tobe. 2017. 'A revised taxonomy of the Felidae: The final report of the Cat Classification Task Force of the IUCN/SSC Cat Specialist Group', *CATnews* Special Issue 11: 1–79.

Law, G. and Andrew C. Kitchener. 2020. 'Twenty years of the tiger feeding pole: Review and recommendations', *International Zoo Yearbook* 54 (1): 174–90.

Less, E. H., K. E. Lukas, R. Bergl, R. Ball, C. W. Kuhar, S. R. Lavin, M. A. Raghanti, J. Wensvoort, M. A. Willis and P. M. Dennis. 2014. 'Implementing a low-starch biscuit-free diet in zoo gorillas. The impact on health: The effect of diet on gorilla health', *Zoo Biology* 33 (1): 74–80.

Li, Youxu, Ronald R. Swaisgood, Wei Wei, Yonggang Nie, Yibo Hu, Xuyu Yang, Xiaodong Gu and Zejun Zhang. 2017. 'Withered on the stem: Is bamboo a seasonally limiting resource for giant pandas?', *Environmental Science and Pollution Research* 24 (11): 10537–46.

Lieberman, Daniel E., Gail E. Krovitz, Franklin W. Yates, Maureen Devlin and Marisa St. Claire. 2004. 'Effects of food processing on masticatory strain and craniofacial growth in a retrognathic face', *Journal of Human Evolution* 46 (6): 655–77.

Lorenzen, Eline D., Peter Arctander and Hans R. Siegismund. 2008. 'High variation and very low differentiation in wide ranging plains zebra (*Equus quagga*): Insights from mtDNA and microsatellites', *Molecular Ecology* 17 (12): 2812–24.

Luke, D. A., C. H. Tonge and D. J. Reid. 1979. 'Metrical analysis of growth changes in the jaws and teeth of normal, protein deficient and calorie deficient pigs', *Journal of Anatomy* 129 (3): 449–57.

Mallinson, Jeremy J. C. 2003. 'A sustainable future for zoos and their role in wildlife conservation', *Human Dimensions of Wildlife* 8 (1): 59–63.

Medeiros, Denis M., Aaron Plattner, Dianne Jennings and Barbara Stoecker. 2002. 'Bone morphology, strength and density are compromised in iron-deficient rats and exacerbated by calcium restriction', *Journal of Nutrition* 132 (10): 3135–41.

Mehta, G., H. I. Roach, S. Langley-Evans, P. Taylor, I. Reading, R. O. C. Oreffo, A. Aihie-Sayer, N. M. P. Clarke and C. Cooper. 2002. 'Intrauterine exposure to a maternal low protein diet reduces adult bone mass and alters growth plate morphology in rats', *Calcified Tissue International* 71 (6): 493–8.

Miller, Scott C., Bernard P. Halloran, Hector F. DeLuca and Webster S. S. Jee. 1983. 'Studies on the role of vitamin D in early skeletal development, mineralization, and growth in rats', *Calcified Tissue International* 35 (1): 455–60.

Mitchell, D. Rex, Stephen Wroe, Matthew J. Ravosa and Rachel A. Menegaz. 2021. 'More challenging diets sustain feeding performance: Applications toward the captive rearing of wildlife', *Integrative Organismal Biology* 3 (1): 1–13.

Moore, W. J. 1965. 'Masticatory function and skull growth', *Proceedings of the Zoological Society of London* 146 (2): 123–31.

Morales-García, Nuria Melisa, Pamela G. Gill, Christine M. Janis and Emily J. Rayfield. 2021. 'Jaw shape and mechanical advantage are indicative of diet in Mesozoic mammals', *Communications Biology* 4 (1, 242): 1–14.

Neaux, Dimitri, Hugo Harbers, Barbara Blanc, Katia Ortiz, Yann Locatelli, Anthony Herrel, Vincent Debat and Thomas Cucchi. 2022. 'The effect of captivity on craniomandibular and calcaneal ontogenetic trajectories in wild boar', *Journal of Experimental Zoology Part B: Molecular and Developmental Evolution* 338 (8): 575–85.

Nett, Emily M., Brielle Jaglowski, Luca J. Ravosa, Dominick D. Ravosa and Matthew J. Ravosa. 2021. 'Mechanical properties of food and masticatory behavior in llamas, *Llama glama*', *Journal of Mammalogy* 102 (5): 1375–89.

Nijboer, Joeke and Ellen S. Dierenfeld. 1996. 'Comparison of diets fed to southeast Asian colobines in North American and European zoos, with emphasis on temperate browse composition', *Zoo Biology* 15 (5): 499–507.

Olmedilla, B. and F. Granado. 2000. 'Growth and micronutrient needs of adolescents', *European Journal of Clinical Nutrition* 54 (S1): S11–15.

O'Regan, Hannah J. and Andrew C. Kitchener. 2005. 'The effects of captivity on the morphology of captive, domesticated and feral mammals', *Mammal Review* 35 (3–4): 215–30.

Plowman, Amy. 2013. 'Diet review and change for monkeys at Paignton Zoo Environmental Park', *Journal of Zoo and Aquarium Research* 1 (2): 73–7.

Plowman, Amy and Francis Cabana. 2019. 'Transforming the nutrition of zoo primates (or how we became known as loris man and that evil banana woman)'. In *Scientific Foundations of Zoos and Aquariums*, edited by Allison B. Kaufman, Meredith J. Bashaw and Terry L. Maple, 274–303. Cambridge: Cambridge University Press.

Pokhojaev, Ariel, Hadas Avni, Tatiana Sella-Tunis, Rachel Sarig and Hila May. 2019. 'Changes in human mandibular shape during the Terminal Pleistocene–Holocene Levant', *Scientific Reports* 9 (1, 8799): 1–10.

Ravosa, Matthew J., Jeremiah E. Scott, Kevin R. McAbee, Anna J. Veit and Annika L. Fling. 2015. 'Chewed out: An experimental link between food material properties and repetitive loading of the masticatory apparatus in mammals', *PeerJ* 3 (e1345): 1–19.

Regaiolli, Barbara, Chiara Angelosante, Giovanna Marliani, Pier Attilio Accorsi, Stefano Vaglio and Caterina Spiezio. 2020. 'Gum feeder as environmental enrichment for zoo marmosets and tamarins', *Zoo Biology* 39 (2): 73–82.

Roe, Katie, Andrew McConney and Caroline F. Mansfield. 2014. 'The role of zoos in modern society: A comparison of zoos' reported priorities and what visitors believe they should be', *Anthrozoös* 27 (4): 529–41.

Ross, Stephen R. and Marisa A. Shender. 2016. 'Daily travel distances of zoo-housed chimpanzees and gorillas: Implications for welfare assessments and space requirements', *Primates* 57 (3): 395–401.

Rude, Robert K., H. E. Gruber, L. Y. Wei, A. Frausto and B. G. Mills. 2003. 'Magnesium deficiency: Effect on bone and mineral metabolism in the mouse', *Calcified Tissue International* 72 (1): 32–41.

Rude, Robert K., Mary E. Kirchen, Helen E. Gruber, Audrey A. Stasky and Martha H. Meyer. 1998. 'Magnesium deficiency induces bone loss in the rat', *Mineral and Electrolyte Metabolism* 24 (5): 314–20.

Ruff, Christopher, Brigette Holt and Erik Trinkaus. 2006. 'Who's afraid of the big bad Wolff?: "Wolff's law" and bone functional adaptation', *American Journal of Physical Anthropology* 129 (4): 484–98.

Savarino, Giovanni, Antonio Corsello and Giovanni Corsello. 2021. 'Macronutrient balance and micronutrient amounts through growth and development', *Italian Journal of Pediatrics* 47 (1, 109): 1–14.

Schulte-Hostedde, Albrecht I. and Gabriela F. Mastromonaco. 2015. 'Integrating evolution in the management of captive zoo populations', *Evolutionary Applications* 8 (5): 413–22.

Searcy, William A., Susan Peters and Stephen Nowicki. 2004. 'Effects of early nutrition on growth rate and adult size in song sparrows *Melospiza melodia*', *Journal of Avian Biology* 35 (3): 269–79.

Shi, Jun-shuai, Rui Gu, Shuang-lin Chen, Chao Zhang and Zi-wu Guo. 2019. 'The effect of altitude on the protein nutritional value of *Phyllostachys prominens* bamboo shoots', *Acta Agriculturae Universitatis Jiangxiensis* 41 (2): 308–15.

Siciliano-Martina, Leila, Jessica E. Light and A. Michelle Lawing. 2021. 'Cranial morphology of captive mammals: A meta-analysis', *Frontiers in Zoology* 18 (1): 1–13.

Siciliano-Martina, Leila, Margot Michaud, Brian P. Tanis, Emily L. Scicluna and A. Michelle Lawing. 2022. 'Endocranial volume increases across captive generations in the endangered Mexican wolf', *Scientific Reports* 12 (1, 8147): 1–8.

Silva, Matthew J., Jeremy D. Eekhoff, Tarpit Patel, Jane P. Kenney-Hunt, Michael D. Brodt, Karen Steger-May, Erica L. Scheller and James M. Cheverud. 2019. 'Effects of high-fat diet and body mass on bone morphology and mechanical properties in 1100 advanced intercross mice', *Journal of Bone and Mineral Research* 34 (4): 711–25.

Stuhlträger, Julia, Ellen Schulz-Kornas, Roman M. Wittig and Kornelius Kupczik. 2019. 'Ontogenetic dietary shifts and microscopic tooth wear in western chimpanzees', *Frontiers in Ecology and Evolution* 7 (298): 1–14.

Tan, C. L. and J. H. Drake. 2001. 'Evidence of tree gouging and exudate eating in pygmy slow lorises (*Nycticebus pygmaeus*)', *Folia Primatologica* 72 (1): 37–9.

Teaford, Mark F., P. W. Lucas, Peter S. Ungar and Ken Glander. 2006. 'Mechanical defenses in leaves eaten by Costa Rican howling monkeys (*Alouatta palliata*)', *American Journal of Physical Anthropology* 129 (1): 99–104.

Thexton, A. J., K. M. Hiiemae and A. W. Crompton. 1980. 'Food consistency and bite size as regulators of jaw movement during feeding in the cat', *Journal of Neurophysiology* 44 (3): 456–74.

Thompson, C. L., M. M. Valença-Montenegro, L. C. d. O. Melo, Y. B. M. Valle, M. A. B. d. Oliveira, P. W. Lucas and C. J. Vinyard. 2014. 'Accessing foods can exert multiple distinct, and potentially competing, selective pressures on feeding in common marmoset monkeys', *Journal of Zoology* 294 (3): 161–9.

Tribe, Andrew and Rosemary Booth. 2003. 'Assessing the role of zoos in wildlife conservation', *Human Dimensions of Wildlife* 8 (1): 65–74.

Turvey, Samuel T., Jennifer J. Crees, Zhipeng Li, Jon Bielby and Jing Yuan. 2017. 'Long-term archives reveal shifting extinction selectivity in China's postglacial mammal fauna', *Proceedings of the Royal Society B: Biological Sciences* 284 (20171979): 1–10.

Vester, Brittany M., Alison N. Beloshapka, Ingmar S. Middelbos, Sarah L. Burke, Cheryl L. Dikeman, Lee G. Simmons and Kelly S. Swanson. 2009. 'Evaluation of nutrient digestibility and fecal characteristics of exotic felids fed horse- or beef-based diets: Use of the domestic cat as a model for exotic felids', *Zoo Biology* 29 (4): 432–48.

Washburn, S. L. 1947. 'The relation of the temporal muscle to the form of the skull', *Anatomical Record* 99 (3): 239–48.

Watt, D. G. and C. H. M. Williams. 1951. 'The effects of the physical consistency of food on the growth and development of the mandible and the maxilla of the rat', *American Journal of Orthodontics* 37 (12): 895–928.

West, Annie G. and Carolyn M. King. 2018. 'Variation in mandible shape and body size of house mice *Mus musculus* in five separate New Zealand forest habitats', *New Zealand Journal of Zoology* 45 (2): 136–53.

Wozencraft, W.C. 2005. 'Order carnivora'. In *Mammal Species of the World: A Taxonomic and Geographic Reference*, 3rd edition, edited by Don E. Wilson and DeeAnn M. Reeder, 532–628. Baltimore: Johns Hopkins University Press.

Wroe, Stephen. 2007. 'Cranial mechanics compared in extinct marsupial and extant African lions using a finite-element approach', *Journal of Zoology* 274 (4): 332–9.

Wroe, Stephen, Toni L. Ferrara, Colin R. McHenry, Darren Curnoe and Uphar Chamoli. 2010. 'The craniomandibular mechanics of being human', *Proceedings of the Royal Society B: Biological Sciences* 277 (1700): 3579–86.

Yamamoto, Saburo. 1996. 'The effects of food consistency on maxillary growth in rats', *European Journal of Orthodontics* 18 (1): 601–15.

Zernicke, R. F., G. J. Salem, R. J. Barnard and E. Schramm. 1995. 'Long-term, high-fat-sucrose diet alters rat femoral neck and vertebral morphology, bone mineral content, and mechanical properties', *Bone* 16 (1): 25–31.

Zhang, Baowei, Ming Li, Zejun Zhang, Benoît Goossens, Lifeng Zhu, Shanning Zhang, Jinchu Hu, Michael W. Bruford and Fuwen Wei. 2007. 'Genetic viability and population history of the giant panda, putting an end to the "evolutionary dead end"?', *Molecular Biology and Evolution* 24 (8): 1801–10.

Zhao, Feng, Fei Du, Hadrien Oliveri, Lüwen Zhou, Olivier Ali, Wenqian Chen, Shiliang Feng, Qingqing Wang, Shouqin Lü, Mian Long, Rene Schneider, Arun Sampathkumar, Christophe Godin, Jan Traas and Yuling Jiao. 2020. 'Microtubule-mediated wall anisotropy contributes to leaf blade flattening', *Current Biology* 30 (20): 3972–85.

Zhu, LiFeng, YiBo Hu, ZeJun Zhang and FuWen Wei. 2013. 'Effect of China's rapid development on its iconic giant panda', *Chinese Science Bulletin* 58 (18): 2134–9.

10
The effects of red fox chronic exposure to metals on health and the environment
Blessing Chidimuro

Introduction

There is a growing global recognition of the threat posed by toxic heavy metals to animals and humans. Like many other industrialised countries, Britain has seen a change in heavy metal emissions over the past few decades. Heavy metal contamination of the British environment has been and continues to be a problem for animal, human and environmental health (Davies 1997). Metals occur naturally in the environment, but human activity increases their concentrations beyond the natural baseline. Urbanisation, industrialisation, motorisation and agriculture (the latter through fertilisers and sewage sludge) have led to increased metals in the environment, some of them extremely toxic to animals (Nriagu 1988; Järup 2003). Environmental contamination with heavy metals has severe consequences for animal health and the environment (Alonso et al. 2000; Newth et al. 2016; Kalisińska 2019). Animals can be exposed to heavy metals in several ways, including inhalation of contaminated air, consumption of contaminated food, and ingestion of contaminated soil and water (Patra et al. 2011). The transfer of toxic pollutants to animals is influenced by various factors, such as the type of food consumed, as this can impact how certain metals are absorbed by an organism (Hunter et al. 1987).

Heavy metal contamination in animals is often more prevalent in urban areas due to the abundance of human food waste. This waste attracts animals to urban environments, where they scavenge. Human food waste often contains toxic heavy metals, and for some animals, this has become a significant part of their diet. This has led to an increase in opportunistic urban animal species, such as red foxes, red kites (discussed further in

Chapters 4 and 11), corvids (Chapter 5), and pigeons (Chapter 6), as they supplement their diets with human food waste.

In addition to consuming human food waste, animals in urban areas are also at risk of exposure to heavy-metal-contaminated food due to the intentional or unintentional feeding of wildlife by humans. As observed throughout this book and highlighted especially for urban species in Chapters 4 and 6, feeding of animals by humans is widespread. People in urban areas often discard or intentionally provide food items such as raw fish, edible roots and vegetables, which can expose wildlife to heavy metal contamination. It is worth noting that the consequences of animals being exposed to heavy metal contamination due to human activities are far-reaching, and the implications for wildlife populations in urban environments are significant. Knowledge of toxic metal concentration in animals is therefore essential for assessing the polluting effects of human activity on animals.

Until recently, most of the studies concerning the accumulation and effects of heavy metals in wild terrestrial mammals were carried out on rodents (Sawicka-Kapusta et al. 1994; Gdula-Argasińska et al. 2004; Janiga et al. 2019). However, because of their increased utilisation of urbanised habitats, red foxes (*Vulpes vulpes*) may be more prone to exposure to pollutants. Urban habitats are defined as any area characterised by a dense human population and where the land is associated with high concentrations of buildings and infrastructure (Baker and Harris 2007; Adams 2016). Currently, the impact of anthropogenic metal pollutants on British red foxes is unclear, as this has not been studied in great depth. My study examined the levels of arsenic (As), cadmium (Cd), chromium (Cr), and lead (Pb) exposure in red foxes collected in London during the 1970s and compared them to those collected in the early 2020s. The objective was to understand the impact of human activities, such as intentional or unintentional feeding of animals, on the animals' exposure to environmental pollutants. The study also sought to shed light on the potential risks posed by human behaviour on red foxes' health.

The red fox (*Vulpes vulpes*)

The red fox is a widely distributed terrestrial mammal within Britain. The first documentation of red foxes dwelling within urban London is in the 1930s, although they may have been present much earlier (Teagle 1967; Harris 1995). Evidence reveals that the red fox population in Britain was kept low during the eighteenth and early-nineteenth centuries due to fox control in rural areas by people who depended on rabbits and other game animals for their meat, as well as to the popular sport of fox hunting (Lloyd 1980). However, it has been reported that by the end of the nineteenth century the British started importing them from continental Europe to sustain fox hunting (Vesey-Fitzgerald 1965). Since then, they have increased substantially in twentieth- and twenty-first-century London, with their population currently estimated at ten thousand by the London Wildlife Trust (Teagle 1967; Harris 1995; Antony 2022).

Red foxes are extremely adaptable animals, which can utilise a variety of habitats and survive in a range of environments including areas with high exposure to humans such as towns and cities. They are primarily carnivorous, with diverse diets including carrion, small mammals, birds, insects, earthworms, fruits, plants and human food waste (Castañeda et al. 2022). Human–fox interactions are increasing in Britain, with reports showing that foxes have become bolder and are no longer scared of being in close proximity to humans, with some of them living and breeding in people's back gardens (Cassidy and Mills 2012; Padovani et al. 2021).

Red foxes have benefited from their close proximity to humans, as it provides them with easier access to food (Baumann et al. 2020). There is evidence from archaeological findings showing that red foxes have coexisted commensally with humans for a long time, with their proximity leading to them being regularly hunted and bred in some societies for their fur, meat, and teeth (Yeshurun et al. 2009; Conard et al. 2013; Camarós et al. 2016; Baumann et al. 2020). This contrasts quite distinctly with complex modern relationships between humans and foxes in the UK, where products and utilitarian outputs are not the primary focus of interactions with foxes, and attitudes to – and treatment of – foxes often conflict. Similar to what has been observed for red kites (Chapter 4), corvids (Chapter 5) and feral pigeons (Chapter 6), red foxes are often persecuted by humans by intentionally *preventing* their feeding. Traditions such as fox hunting, the promoted use of fox fur for clothing to encourage culling and the negative portrayal of foxes in literature and film (Diaz-Ruiz et al. 2015) mean that many people view red foxes as predators and pests. In direct opposition to this, many other people intentionally feed foxes in their gardens (Baker et al. 2004), engaging in non-utilitarian and affective relationships with these animals, which ultimately reinforce their proximity with us and our food waste.

In recent years, intentional feeding of anthropogenic food to red foxes has increased significantly in urban areas, mainly due to urban expansion driven by population growth (Bateman and Fleming 2012;

Baker and Harris 2007). Previous studies have shown that red foxes are adaptable in their diet, and their food choices depend on availability and accessibility (Macdonald 1989; Cavallini and Volpi 1995; Leckie et al. 1998). In urban areas, red foxes scavenge a wide variety of anthropogenic food including bones, carcasses, meat, bread and pet food often left by humans in their gardens (Baker et al. 2004). Their opportunistic feeding behaviour has led them to adapt to consuming human garbage. As a result, it can be assumed that the surplus supply of anthropogenic food is most likely responsible for an increase in the number of foxes in urban areas. High pollution is also often found in cities; therefore, the increased population of foxes in these cities exposes them to these contaminants. Red foxes' increased utilisation of urbanised and agricultural habitats potentially exposes them to metal pollutants that, in turn, may adversely affect their health. Despite the threat of metal pollutants to red foxes, no studies have previously been devoted to this issue in Britain.

The pollutants

Lead (Pb)

Lead is a chemical element that is one of the most abundant toxic metals in the environment. It is mainly found in industrial regions, with common sources including lead-based paints, petrol, batteries, and combustion of coal and mineral oil (Patra et al. 2007). In the UK, the deposition of lead in the atmosphere increased from the start of the Industrial Revolution in the eighteenth century. During the twentieth century, vehicle exhaust emissions constituted a major source of lead pollution through the use of leaded petrol, as well as the use of lead-based paint (Harrison and Johnston 1985; O'Brien and Roberts 2011). There have been large decreases in lead concentrations in the British environment since the 1970s, reflecting increased use of unleaded petrol, disuse of leaded paints and decreasing use of coal. Records show that between 1970 and 2008, lead emissions declined by 99 per cent (Murrells et al. 2010).

Although lead emissions fell due to its removal from petrol (O'Brien and Roberts 2011), household paint (UK statutory Instruments 1992), solder and other consumer products, lead concentrations in the air remain high in large cities (Resongles et al. 2021). Animals are exposed to lead mainly through the intake of carrion and game animals shot by lead bullets, contaminated drinking water and inhalation of contaminated air. Additionally, lead exposure can result from contamination of feed and soil from industrial pollution and agricultural practices (Patra et al. 2011). Lead can become harmful to animals if it accumulates in tissues after prolonged exposure, even in small quantities (Ercal et al. 2001). Exposure to lead has been known to affect the brain tissue of foxes, resulting in impaired motor skills, convulsions and increased aggression (Pattee and Pain 2002; Kalisińska et al. 2023).

Cadmium (Cd)

Similarly to lead, cadmium is one of the most toxic metals in the environment, where it is present due to human activities such as mining, agriculture and industry (Genchi et al. 2020). Cadmium has accumulated in the environment mainly as a consequence of fossil fuel combustion, copper and nickel smelting and refining in mining, disposal of urban refuse and from the use of phosphate fertilisers in agriculture (Haider et al. 2021). It is also used in the manufacturing of nickel-cadmium (Ni-Cd) batteries, as a stabiliser in PVC products, as a corrosive reagent and as a pigment (Pandey and Sharma 2014; Genchi et al. 2020). Cadmium emissions have declined by 92 per cent since 1970, mainly because of the closure of coal mines and the decline in fuel oil combustion in power generation (Murrells et al. 2010). Additionally, due to its presence in commercial phosphate fertilisers, proposals made in 2002 to limit its concentrations in the EU have led to the reduction of cadmium in agricultural soils across Europe (Smolders and Six 2013).

The exposure to animals primarily occurs through the ingestion of cadmium-contaminated crops. Its availability in phosphate fertilisers makes it necessary for it to be absorbed by plants, resulting in plants often having higher cadmium concentrations compared to meat products (Pandey and Sharma 2014; Satarug 2018). Vegetables, cereal sub-products, starchy roots, nuts and pulses contribute to cadmium dietary intake, thereby accounting for high concentrations in animals that ingest them (or in animals that consume other animals that in turn feed on these plants). It can also be absorbed into animals' bodies through the air (Satarug 2019). Cadmium is recognised as a potential health threat to animals (Scheuhammer 1987); it has been shown to affect skeletal integrity and calcium balance in wild animals, with the liver and kidneys mostly affected (Friberg et al. 1986; Genchi et al. 2020). A high cadmium diet can cause a reduction in the uptake of calcium in animals, leading to bone defects (Ando et al. 1978; Pandey and Sharma 2014).

Arsenic (As)

Arsenic is one of the commonest environmental pollutants in the world. Although it is found naturally in rocks, sediments and soils, human activities – such as agriculture, mining, smelting, glass manufacturing, coal combustion, and industrial and domestic waste disposal – contribute to its high prevalence in the environment (Bosch et al. 2016; Kesici 2016). Its toxicity to animals is related to its chemical form, with its inorganic forms (arsenite and arsenate) being more toxic compared to its organic forms (methylarsenate and dimethylarsinate; Ventura-Lima et al. 2011). Mining operations and metal smelting in particular have led to high concentrations of arsenic in urban areas (Kesici 2016). Currently, high concentrations of arsenic in soils are derived from mining and emissions from industries, such as pigment manufacture, electronics, cosmetics, ceramics and glass (Nriagu et al. 2007). Additionally, it is also acquired through the use of inorganic arsenic herbicides, pesticides, fertilisers and desiccants in agriculture (Liu et al. 2015). Furthermore, due to its natural presence in rocks, erosion can result in arsenic being released into lakes and rivers, and therefore drinking water. It may additionally enter the water through industrial and waste disposal (Jang et al. 2016).

Animal exposure to arsenic occurs through inhalation of air and dust, ingestion of contaminated food and water, and direct contact with soil (Drouhot et al. 2014). Arsenic in food is mostly found in seafood, seaweed, poultry, rice and mushrooms (Bosch et al. 2016; Kesici 2016). High concentrations of arsenic have been known to affect almost every bodily system, including respiratory, cardiovascular, nervous, reproductive and immune systems in animals (Kahn and Line 2008). Records indicate that arsenic emissions have been reduced by 83 per cent since 1970 due to the decline of coal combustion in Britain, with stringent control measures in place to minimise its contamination of air, water and food to the lowest practical level (Murrells et al. 2010; Public Health England 2019).

Chromium (Cr)

Chromium is another toxic heavy metal that is highly dangerous to animals. It is released into the environment through industrial operations such as coal combustion, chrome plating, metal refining, metallurgy, textile dyes and electroplating. Its introduction to the environment is mostly through improper disposal of chromium-contaminated waste from these industries (Mitra et al. 2017; Coetzee et al. 2020). It is considered to be one of the most prominent pollutants in groundwater and soil as it can easily infiltrate them through waste disposal and other natural sources (Kośla et al. 2019).

Animals are exposed to chromium through inhalation of contaminated air, skin contact and ingestion of contaminated food and water (Kośla et al. 2019). High concentrations of chromium affect most systems in an animal body, including the respiratory tract, skin, gastrointestinal, renal, reproductive and cardiovascular systems. It can also cause cancers and tissue damage (Teklay 2016). Fish have been found to have the highest concentrations of chromium, with high concentrations also observed in vegetables, fruits, meat and cereals (Kapoor et al. 2022). Chromium emissions in Britain have reduced by 88 per cent since 1970 due to a reduction in the use of coal in industry, as well as the treatment of chromium-contaminated waste before it enters groundwater and soil (Bewley 2007; Murrells et al. 2010).

Materials and methods

A total of 243 red foxes were analysed, all collected from London and the surrounding countryside and held in the collection of National Museums Scotland (Edinburgh). Two time periods were represented: 1971 to 1973 (n = 129, collected by Steve Harris, referred to here onwards as the 1970s foxes; Harris 1975) and 2021 to 2022 (n = 114, collected by veterinarians and the London Fox Project, referred to as the 2020s foxes). The foxes had died through various causes, including most commonly road traffic accidents and killing by pest controllers.

Previous studies have measured lead concentrations in animal bones using K-shell X-ray fluorescence (KXRF) and inductively coupled plasma mass spectrometry (ICP-MS; see Wallens 2018; Specht et al. 2019a; Tajchman et al. 2020). However, these methods are expensive, time-consuming, not portable and can cause damage to archived specimens. Portable X-ray fluorescence (pXRF), on the other hand, is a non-destructive and portable method for identifying a wide range of elements simultaneously, using a handheld machine that directs X-rays towards the material being studied, resulting in the emission of fluorescent X-rays that are specific to different elements. This method has previously been used to measure lead in bird bones (Specht et al. 2018; Specht et al. 2019a and 2019b; Hampton et al. 2021).

The pXRF measurements of the fox skulls were taken using a ThermoFisher Niton XL3 Portable XRF Analyser, recording a total of about fifty elements, including concentrations of lead, cadmium, chromium and arsenic (expressed in moles). Statistical analysis and data visualisation were carried out using PAST (Hammer et al. 2001) and IBM SPSS (version 26).

Results and Discussion

The results for all the metallic contaminants described within the two fox populations above are given in Figure 10.1.

The studied foxes manifested a considerable range in lead and chromium concentrations, with distinctively higher concentrations in the 1970s than in the 2020s London foxes (Mann-Whitney U tests: Pb U = 5294.5, p < 0.05; Cr U = 3665, p < 0.05). This may be connected

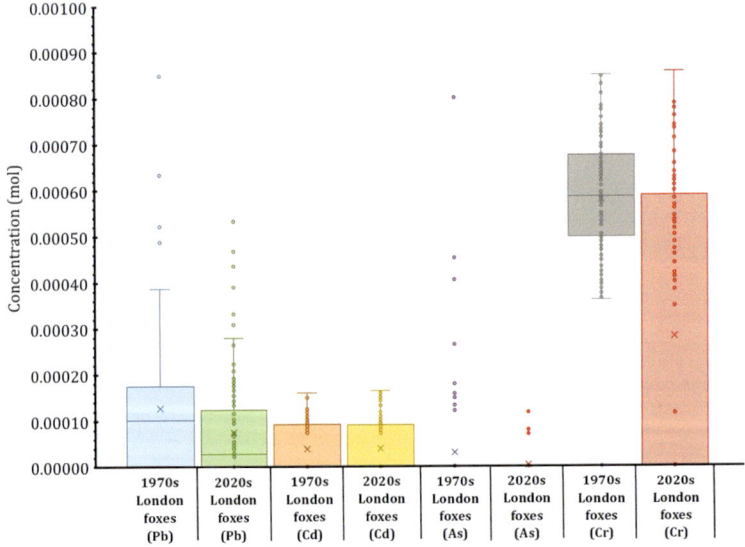

Figure 10.1 Boxplots showing the comparisons of heavy metal concentrations – lead (Pb), cadmium (Cd), arsenic (As) and chromium (Cr) – in red foxes in London between the 1970s and the 2020s. The absence of quartile lines in boxplots for some metals occurs when the quartiles are equal to each other. Lead and chromium are higher in the 1970s while cadmium is similar in both groups of foxes. For arsenic, although the outlier results were lower in the 2020s, the median and quartile measurements were recorded as zero in both groups of foxes. *Source*: author.

with the prevalent use of leaded petrol, lead paint, and coal and mineral oil combustion in the 1970s (Murrells et al. 2010). Despite the reduction measures implemented to reduce all these metal concentrations in the environment, the results show that some 2020s foxes still have high lead and chromium concentrations. Lead emitted from vehicle brakes, tyres and metal equipment from industrial areas, lead-contamination of sewage sludge as well as lead-contaminated air from historical leaded petrol usage have been identified as sources of lead in London today (Dong et al. 2017; Resongles et al. 2021). Resongles et al.'s (2021) study of airborne particles in London between 2014 and 2018 revealed that there is still a high lead concentration as a result of leaded petrol and the concentration has not significantly changed since its ban in 2000; the pXRF results demonstrate this is a health concern for both human and non-human residents (Ercal et al. 2001).

The extensive use of sewage sludge as a fertiliser in London (Hough et al. 2018; James Hutton Institute 2021) may additionally be contributing to the high lead concentration in some of these modern London foxes. It has been established that heavy metals such as lead tend to accrue in sewage sludge and accumulate in plants grown on contaminated land, which in turn accumulate up the food chain (Alloway 2013). Such high concentrations could also result from foxes ingesting lead or having been shot with lead ammunition; some of the foxes analysed in this study had been killed by lead pellets. It must also be considered that the foxes may have eaten prey shot with lead bullets or a diet contaminated with lead, such as earthworms and other soil invertebrates, which often contain high levels of lead (Hopkin 1989). The localisation of the 2020s foxes from highly urbanised areas of London indicates lead contamination from eating within the city. As modern foxes consume greater quantities of human food waste, they may also be eating food contaminated with lead imported from countries with fewer regulations surrounding metallic contaminants.

On the other hand, the 1970s London foxes' arsenic and cadmium means are not statistically different to those of the 2020s London foxes (in the Mann-Whitney U tests I conducted: As $U = 6726.5$, $p = 0.390$; Cd $U = 7339$, $p = 1.000$). The high cadmium and arsenic values in the 2020s foxes may be associated with more coastal influences and feeding on contaminated fish (Kapoor et al. 2022); this would further demonstrate how dangerously widespread metallic contaminants are within the food web as well as in human food sources.

Despite the reduction in cadmium and arsenic concentrations in the environment from the 1970s to the present, it has been observed in this study that the 1970s mean concentrations fall within the range found in 2020s foxes, suggesting similar exposure to these metals. The presence of these metals in the 1970s foxes is expected as their concentration in the environment was higher during that time.

Significant correlations – a measure of the extent to which two random variables change together, in this case whereby when one pollutant increases, the other also increases, and vice versa – were observed among chromium and cadmium concentrations in the 1970s foxes and lead and arsenic concentrations in the 2020s foxes. These significant correlations indicate common sources, probably represented in the foxes' regular dietary items. On the contrary, correlations between chromium and cadmium on the one hand, and between lead and arsenic on the other, were not found in the 1970s and 2020s foxes respectively, suggesting that there are separate sources of contamination for these two populations. High levels of contamination in the environment where they are living is the most plausible explanation for this scenario. Heavy metal concentrations are expected to increase with proximity to urban areas, areas often associated with concentrated anthropogenic pollutants. Differences in habitat might thus lead to differences in metal exposure and accumulation in these foxes. There seems to be no relationship between the location from which both populations were collected and the metal concentrations in them. Although these foxes were collected from a wide range of locations, these were most likely not the original habitats in which they mostly lived. Across all environments these animals are continuously exposed to a range of environmental contaminants, including heavy metals, though in lower concentrations in rural settings. The concentrations of these metals found within an individual fox can serve as a valuable indicator of its habitat type – urban or rural. Urban foxes typically encounter a higher concentration of pollutants due to the significant amount of human activity and industrial processes prevalent in city settings. These pollutants often arise from sources such as traffic emissions, waste disposal, and construction activities, leading to increased levels of heavy metals in the soil and food sources available to these animals. In contrast, rural foxes generally inhabit areas with lower levels of industrialisation and human interference, resulting in reduced exposure to these harmful substances.

As a result, analysing heavy metal concentrations in red foxes can provide critical insights into the environmental health of different habitats and the potential risks associated with urban living for wildlife. Understanding these differences not only aids in wildlife management efforts but also raises awareness about the broader implications of urban pollution on ecosystem health (Chidimuro et al. in prep.).

It is notable that the concentrations seen in both the 1970s and 2020s foxes go far beyond baselines for heavy metal contamination resulting from the areas where they lived. While the decrease in metallic contaminants between the populations is seen to mark success in reducing environmental pollution, it is evident that modern urban foxes still intake a great quantity of toxic metals within their diet. It therefore becomes clear that while the abundance of anthropogenic foods is what has drawn red foxes closer to humans, it is these same food sources that are damaging their health, including reductions in fertility and possible increases in aggression due to lead poisoning. Unintentional – or in some cases, intentional (Varela 2018) – feeding of foxes by humans is, therefore, crucial to explore through further research, considering the ways in which our waste is harming the creatures we share environments with.

Conclusion

In conclusion, it was shown that the foxes accumulated a different mixture of heavy metals despite being collected from the same city fifty years apart. The increased metal concentrations detected in some of the foxes compared to others within the same population support the conclusion that they may be occupying different habitats or exploiting different food sources. The study established a strong link between increased human activities – especially the intentional and/or unintentional feeding of foxes – and animals' exposure to environmental pollution. This study suggests that red foxes sharing the same environment and food with humans show increased concentrations of heavy metals and therefore most likely were living in urban areas. Heavy metal concentrations in tissues thereby constitute a valid marker for the classification of red foxes as urban in terms of their habitat. Additionally, the study shows the importance and validity of reducing heavy metals in the environment as evidenced by the lower concentrations in the 2020s foxes compared to the 1970s foxes. Moreover, the results obtained in this study underline the importance of red foxes being valid bioindicators of environmental pollution by heavy metals.

Overall, this study highlights that the occurrence of high heavy-metal concentrations in red foxes is associated with human activities. It has been observed that humans have a big impact on both behavioural and physiological changes of red foxes, and that direct and indirect contact between our species is fundamentally dependent on food availability, whether intentionally or unintentionally supplied. As foxes continue to live close to humans, becoming bolder and continuing to eat human food waste, humans are actively creating animal health concerns due to improper management of food waste.

References

Adams, Clarke E. 2016. *Urban Wildlife Management*. Boca Raton, FL: CRC Press.

Alloway, Brian J. 2013. 'Sources of heavy metals and metalloids in soils'. In *Heavy Metals in Soils: Trace metals and metalloids in soils and their bioavailability*, edited by Brian J. Alloway, 11–50. Dordrecht: Springer Netherlands.

Alonso, M. López, J. L. Benedito, M. Miranda, C. Castillo, J. Hernández and R. F. Shore. 2000. 'Arsenic, cadmium, lead, copper and zinc in cattle from Galicia, NW Spain', *Science of the Total Environment* 246 (2–3): 237–48.

Ando, Masanori, Yasuyoshi Sayato and Toshiaki Osawa. 1978. 'Studies on the disposition of calcium in bones of rats after continuous oral administration of cadmium', *Toxicology and Applied Pharmacology* 46 (3): 625–32.

Antony, Andrew. 2022. 'Urban foxes: Are they "fantastic" or a growing menace?'. *The Guardian*. Accessed 16 May 2023. https://www.theguardian.com/environment/2022/oct/15/urban-foxes-are-they-fantastic-or-a-growing-menace.

Baker, Philip J., Stephan Michael Funk, Stephen A. Harris, Tabetha J. Newman, Glen R. Saunders and Piran C. L. White. 2004. 'The impact of human attitudes on the social and spatial organization of urban foxes (*Vulpes vulpes*) before and after an outbreak of sarcoptic mange'. In *Proceedings of the 4th International Symposium on Urban Wildlife Conservation*, edited by William W. Shaw, Lisa K. Harris and Larry Vandruff, 153–63. Tucson: University of Arizona.

Baker, Philip J. and Stephen Harris. 2007. 'Urban mammals: What does the future hold? An analysis of the factors affecting patterns of use of residential gardens in Great Britain', *Mammal Review* 37 (4): 297–315.

Bateman, Philip W. and Patricia Anne Fleming. 2012. 'Big city life: Carnivores in urban environments', *Journal of Zoology* 287 (1): 1–23.

Baumann, Chris, Gillian L. Wong, Britt M. Starkovich, Susanne C. Münzel and Nicholas J. Conard. 2020. 'The role of foxes in the Palaeolithic economies of the Swabian Jura (Germany)', *Archaeological and Anthropological Sciences* 12 (208): 1–17.

Bewley, Richard. 2007. 'Treatment of chromium contamination and chromium ore processing residue'. Accessed 16 May 2023. http://www.mcilvainecompany.com/Decision_Tree/subscriber/Tree/DescriptionTextLinks/ChromiumContamination.pdf.

Bosch, Adina C., Bernadette O'Neill, Gunnar O. Sigge, Sven E. Kerwath and Louwrens C. Hoffman. 2016. 'Heavy metals in marine fish meat and consumer health: A review', *Journal of the Science of Food and Agriculture* 96 (1): 32–48.

Camarós, Edgard, Susanne C. Münzel, Marián Cueto, Florent Rivals and Nicholas J. Conard. 2016. 'The evolution of Paleolithic hominin–carnivore interaction written in teeth: Stories from the Swabian Jura (Germany)', *Journal of Archaeological Science: Reports* 6: 798–809.

Cassidy, Angela and Brett Mills. 2012. '"Fox Tots Attack Shock": Urban foxes, mass media and boundary-breaching', *Environmental Communication* 6 (4): 494–511.

Castañeda, Irene, Tim S. Doherty, Patricia A. Fleming, Alyson M. Stobo-Wilson, John C. Z. Woinarski and Thomas M. Newsome. 2022. 'Variation in red fox *Vulpes vulpes* diet in five continents', *Mammal Review* 52 (3): 328–42.

Cavallini, Paolo and Teresa Volpi. 1995. 'Biases in the analysis of the diet of the red fox *Vulpes vulpes*', *Wildlife Biology* 1 (1): 243–8.

Chidimuro, Blessing, David Cooper, Andrew Kitchener, Steven Harris and Stuart Black. In prep. 'Defining "urban" based on dietary and pollutant signatures in the red fox (*Vulpes vulpes*)'.

Coetzee, Johan J., Neetu Bansal and Evans M. N. Chirwa. 2020. 'Chromium in environment, its toxic effect from chromite-mining and ferrochrome industries, and its possible bioremediation', *Exposure and Health* 12 (1): 51–62.

Conard, Nicholas J., Keiko Kitagawa, Petra Krönneck, Madelaine Böhme and Susanne C. Münzel. 2013. 'The importance of fish, fowl and small mammals in the paleolithic diet of the Swabian Jura, southwestern Germany'. In *Zooarchaeology and Modern Human Origins: Vertebrate paleobiology and paleoanthropology*, edited by Jamie L. Clark and John D. Speth, 173–90. Dordrecht: Springer Netherlands.

Davies, B. E. 1997. 'Heavy metal contaminated soils in an old industrial area of Wales, Great Britain: Source identification through statistical data interpretation', *Water, Air, and Soil Pollution* 94: 85–98.

Diaz-Ruiz, F., J. Caro, M. Delibes-Mateos, B. Arroyo and P. Ferreras. 2015. 'Drivers of red fox (*Vulpes vulpes*) daily activity: Prey availability, human disturbance or habitat structure?', *Journal of Zoology* 298 (2): 128–38.

Dong, Shuofei, Raquel Ochoa Gonzalez, Roy M. Harrison, David Green, Robin North, Geoff Fowler and Dominik Weiss. 2017. 'Isotopic signatures suggest important contributions from recycled gasoline, road dust and non-exhaust traffic sources for copper, zinc and lead in PM10 in London, United Kingdom', *Atmospheric Environment* 165: 88–98.

Drouhot, Séverine, Francis Raoul, Nadia Crini, Christelle Tougard, Anne-Sophie Prudent, Coline Druart, Dominique Rieffel, Jean-Claude Lambert, Nicolas Tête, Patrick Giraudoux and Renaud Scheifler. 2014. 'Responses of wild small mammals to arsenic pollution at a partially remediated mining site in Southern France', *Science of the Total Environment* 470 (1): 1012–22.

Ercal, N., H. Gurer-Orhan and N. Aykin-Burns. 2001. 'Toxic metals and oxidative stress part I: Mechanisms involved in metal-induced oxidative damage', *Current Topics in Medicinal Chemistry* 1 (6): 529–39.

Friberg, Lars, Gunnar Nordberg and Velimir B. Vouk. 1986. *Handbook on the Toxicology of Metals*, 2nd Edition. Volume 1: *General Aspects*. London: Elsevier.

Gdula-Argasińska, Joanna, John Appleton, Katarzyna Sawicka-Kapusta and Bill Spence. 2004. 'Further investigation of the heavy metal content of the teeth of the bank vole as an exposure indicator of environmental pollution in Poland', *Environmental Pollution* 131 (1): 71–9.

Genchi, Giuseppe, Maria Stefania Sinicropi, Graziantonio Lauria, Alessia Carocci and Alessia Catalano. 2020. 'The effects of cadmium toxicity', *International Journal of Environmental Research and Public Health* 17 (11): 1–24.

Haider, Fasih Ullah, Cai Liqun, Jeffrey A. Coulter, Sardar Alam Cheema, Jun Wu, Renzhi Zhang, Ma Wenjun and Muhammad Farooq. 2021. 'Cadmium toxicity in plants: Impacts and remediation strategies', *Ecotoxicology and Environmental Safety* 211 (111887): 1–22.

Hammer, Øyvind, David A. T. Harper and Paul D. Ryan. 2001. 'PAST: Paleontological statistics software package for education and data analysis', *Palaeontologia Electronica* 4 (1): 1–9.

Hampton, Jordan O., Aaron J. Specht, James M. Pay, Mark A. Pokras and Andrew J. Bengsen. 2021. 'Portable X-ray fluorescence for bone lead measurements of Australian eagles', *Science of the Total Environment* 789 (147998): 1–6.

Harris, Stephen. 1975. 'Aspects of the Biology of Suburban Foxes', doctoral thesis, University of London.

Harris, Stephen. 1995. *A Review of British Mammals: Population estimates and conservation status of British mammals other than cetaceans*. Peterborough: JNCC.

Harrison, R. M. and W. R. Johnston. 1985. 'Deposition fluxes of lead, cadmium, copper and polynuclear aromatic hydrocarbons (PAH) on the verges of a major highway', *Science of the Total Environment* 46: 121–35.

Hopkin, Stephen P. 1989. *Ecophysiology of Metals in Terrestrial Invertebrates*. Dordrecht: Kluwer Academic Publishers Group.

Hough, Rupert, David Tompkins, Fiona Nicholson, Steve Pierson, John Williams, Dan Munro and Dominic Duckett. 2018. 'The impacts on human health and environment arising from the spreading of sewage sludge to land (CR/2016/23)'. Scottish Government. Accessed 9 May 2023. https://tinyurl.com/58jax6fp.

Hunter, B. A., M. S. Johnson and D. J. Thompson. 1987. 'Ecotoxicology of copper and cadmium in a contaminated grassland ecosystem. III. Small mammals', *Journal of Applied Ecology* 24: 587–99.

James Hutton Institute. 2021. 'Potentially hazardous agents in land-applied sewage sludge: Human health risk assessment'. Accessed 16 May 2023. https://www.gov.scot/publications/human-health-risk-assessment-potentially-hazardous-agents-land-applied-sewage-sludge/.

Jang, Yong-Chul, Yashoda Somanna and Hwidong Kim. 2016. 'Source, distribution, toxicity and remediation of arsenic in the environment: A review', *Journal of International Environmental Application & Science* 11 (2): 559–81.

Janiga, Marián, Zuzana Ballová, Mária Angelovičová and Ján Korňan. 2019. 'The snow vole and Tatra marmot as different rodent bioindicators of lead pollution in an Alpine environment: A hibernation effect', *Polish Journal of Environmental Studies* 28 (5): 3215–26.

Järup, Lars. 2003. 'Hazards of heavy metal contamination', *British Medical Bulletin* 68: 167–82.

Kahn, Cynthia M. and Scott Line (eds). 2008. *The Merck Veterinary Manual: A handbook of diagnosis, therapy, and disease prevention and control for the veterinarian*. 9th edition. Rahway: Merck & Co.

Kalisińska, Elżbieta. 2019. *Mammals and Birds as Bioindicators of Trace Element Contaminations in Terrestrial Environments: An ecotoxicological assessment of the Northern Hemisphere*. Cham: Springer.

Kalisińska, Elżbieta, Karolina Kot and Natalia Łanocha-Arendarczyk. 2023. 'Red fox as a potential bioindicator of metal contamination in a European environment', *Chemosphere* 319 (138037): 1–11.

Kapoor, Riti Thapar, Manar Fawzi Bani Mfarrej, Pravej Alam, Jörg Rinklebe and Parvaiz Ahmad. 2022. 'Accumulation of chromium in plants and its repercussion in animals and humans', *Environmental Pollution* 301 (119044): 1–11.

Kesici, Gülin Gökçen. 2016. 'Arsenic ototoxicity', *Journal of Otology* 11 (1): 13–17.

Kośla, Tadeusz, Iwona Lasocka and Marta Kołnierzak. 2019. 'Chromium, Cr'. In *Mammals and Birds as Bioindicators of Trace Element Contaminations in Terrestrial Environments: An ecotoxicological assessment of the Northern Hemisphere*, edited by Elżbieta Kalisińska, 57–124. Cham: Springer.

Leckie, F. M., S. J. Thirgood, R. May and S. M. Redpath. 1998. 'Variation in the diet of red foxes on Scottish moorland in relation to prey abundance', *Ecography* 21: 599–604.

Liu, Xueping, Wenfeng Zhang, Yuanan Hu, Erdan Hu, Xiande Xie, Lingling Wang and Hefa Cheng. 2015. 'Arsenic pollution of agricultural soils by concentrated animal feeding operations (CAFOs)', *Chemosphere* 119: 273–81.

Lloyd, H. G. 1980. *The Red Fox*. London: Batsford.

Macdonald, David. 1989. *Running with the Fox*. London: Unwin Hyman.

Mitra, Sayak, Avipsha Sarkar and Shampa Sen. 2017. 'Removal of chromium from industrial effluents using nanotechnology: A review', *Nanotechnology for Environmental Engineering* 2 (11): 1–14.

Murrells, T. P., N. R. Passant, G. Thistlethwaite, A. Wagner, Y. Li, T. Bush, J. Norris, C. Walker, R. A. Stewart, I. Tsagatakis, R. Whiting, C. Conolly, S. Okamura, M. Peirce, S. Sneddon, J. Webb, J. Thomas, J. MacCarthy, S. Choudrie and N. Brophy. 2010. *UK Emissions of Air Pollutants 1970 to 2008*. Didcot: AEA.

Newth, J. L., E. C. Rees, R. L. Cromie, R. A. McDonald, S. Bearhop, D. J. Pain, G. J. Norton, C. Deacon and G. M. Hilton. 2016. 'Widespread exposure to lead affects the body condition of free-living whooper swans Cygnus cygnus wintering in Britain', *Environmental Pollution* 209: 60–7.

Nriagu, J. O. 1988. 'A silent epidemic of environmental metal poisoning?', *Environmental Pollution* 50 (1–2): 139–61.

Nriagu, J. O., P. Bhattacharya, A. B. Mukherjee, J. Bundschuh, R. Zevenhoven and R. H. Loeppert. 2007. 'Arsenic in soil and groundwater: an overview'. In *Trace Metals and other Contaminants in the Environment*, Volume 9, edited by Prosun Bhattacharya, Arun B. Mukherjee, Jochen Bundschuh, Ron Zevenhoven and Richard H. Loeppert, 3–60. Amsterdam: Elsevier.

O'Brien, Elizabeth and Anne Roberts. 2011. 'Chronology of leaded gasoline/leaded petrol history'. Accessed 21 May 2023. https://www.lead.org.au/Chronology-Making_Leaded_Petrol_History.pdf.

Padovani, Roberto, Zhuoyu Shi and Stephen Harris. 2021. 'Are British urban foxes (*Vulpes vulpes*) "bold"? The importance of understanding human–wildlife interactions in urban areas', *Ecology and Evolution* 11 (2): 835–51.

Pandey, Govind and Madhuri Sharma. 2014. 'Heavy metals causing toxicity in animals and fishes', *Research Journal of Animal, Veterinary and Fishery Sciences* 2 (2): 17–23.

Patra, R. C., Amiya K. Rautray and D. Swarup. 2011. 'Oxidative stress in lead and cadmium toxicity and its amelioration', *Veterinary Medicine International* 2011 (457327): 1–9.

Patra, R. C., D. Swarup, Ram Naresh, Puneet Kumar, D. Nandi, Pallav Shekhar, S. Roy and S. L. Ali. 2007. 'Tail hair as an indicator of environmental exposure of cows to lead and cadmium in different industrial areas', *Ecotoxicology and Environmental Safety* 66 (1): 127–31.

Pattee, Oliver H. and Deborah J. Pain. 2002. 'Lead in the environment'. In *Handbook of Ecotoxicology*, 2nd edition, edited by David J. Hoffman, Barnett A. Rattner, G. Allen Burton Jr. and John Cairns Jr., 373–99. Boca Raton, FL: CRC Press.

Public Health England. 2019. 'Arsenic: General information'. Accessed 16 May 2023. https://www.gov.uk/government/publications/arsenic-properties-incident-management-and-toxicology/arsenic-general-information.

Resongles, Eléonore, Volker Dietze, David C. Green, Roy M. Harrison, Raquel Ochoa-Gonzalez, Anja H. Tremper and Dominik J. Weiss. 2021. 'Strong evidence for the continued contribution of lead deposited during the 20th century to the atmospheric environment in London of today', *Proceedings of the National Academy of Sciences* 118 (26, e2102791118): 1–9.

Satarug, Soisungwan. 2018. 'Dietary cadmium intake and its effects on kidneys', *Toxics* 6 (15): 1–23.

Satarug, Soisungwan. 2019. 'Cadmium sources and toxicity', *Toxics* 7 (2): 1–3.

Sawicka-Kapusta, K., Zakrzewska, M. and Chelstowska, J. (1994) 'Heavy metals in small mammals from three different forest areas in Poland', *Acta Biologica Debrecina, Supplementum Oecologica Hungarica* 5: 277–84.

Scheuhammer, A. M. 1987. 'The chronic toxicity of aluminium, cadmium, mercury, and lead in birds: A review', *Environmental Pollution* 46 (4): 263–95.

Smolders, Erik and Laetitia Six. 2013. 'Revisiting and updating the effect of phosphate fertilizers to cadmium accumulation in European agricultural soils'. Accessed 2 May 2023. https://ec.europa.eu/health/scientific_committees/environmental_risks/docs/scher_o_168_rd_en.pdf.

Specht, Aaron J., Chris N. Parish, Emma K. Wallens, Rick T. Watson, Linda H. Nie and Marc G. Weisskopf. 2018. 'Feasibility of a portable X-ray fluorescence device for bone lead measurements of condor bones', *Science of the Total Environment* 615: 398–403.

Specht, Aaron J., Kimberley E. Kirchner, Marc G. Weisskopf and Mark A. Pokras. 2019a. 'Lead exposure biomarkers in the Common Loon', *Science of the Total Environment* 647: 639–44.

Specht, Aaron J., Aisha S. Dickerson and Marc G. Weisskopf. 2019b. 'Comparison of bone lead measured via portable x-ray fluorescence across and within bones', *Environmental Research* 172: 273–8.

Tajchman, Katarzyna, Aleksandra Ukalska-Jaruga, Marek Bogdaszewski, Monika Pecio and Katarzyna Dziki-Michalska. 2020. 'Accumulation of toxic elements in bone and bone marrow of deer living in various ecosystems: A case study of farmed and wild-living deer', *Animals* 10 (11): 1–11.

Teagle, W. G. 1967. 'The fox in the London suburbs', *London Naturalist* 46: 44–68.

Teklay, A. 2016. 'Physiological effect of chromium exposure: A review', *International Journal of Food Science, Nutrition and Dietetics* 7: 1–11.

UK Statutory Instruments. 1992. 'Environmental protection (controls on injurious substances) regulations 1992: Environmental protection'. Accessed 16 May 2023. https://www.legislation.gov.uk/uksi/1992/31/made.

Varela, Charlotte. 2018. 'How to feed foxes and badgers in your garden responsibly'. Accessed 15 February 2024. https://www.lancswt.org.uk/blog/charlotte-varela/how-feed-foxes-badgers-garden.

Ventura-Lima, Juliane, Maurício Reis Bogo and José M. Monserrat. 2011. 'Arsenic toxicity in mammals and aquatic animals: A comparative biochemical approach', *Ecotoxicology and Environmental Safety* 74 (3): 211–10.

Vesey-Fitzgerald, Brian. 1965. *Town Fox, Country Fox*. London: André Deutsch.

Wallens, Emma. 2018. 'Using bone lead as a biomarker for lead toxicity in condors: Accuracy of portable L-X-ray fluorescence (LXRF) device to quantify lead (Pb) in condor bone in vivo', *Journal of Purdue Undergraduate Research* 8: 57–64.

Yeshurun, Reuven, Guy Bar-Oz and Mina Weinstein-Evron. 2009. 'The role of foxes in the Natufian economy', *Before Farming* 1: 1–16.

11
The pros, cons and contrary consequences of conservation feeding: anthropogenic feeding of the red kite (*Milvus milvus*) in Britain

Virginia Thomas

This chapter is dedicated to AG, who knows more about human–animal relations than I ever will.

Introduction

A plethora of terms exists to describe different forms of wildlife feeding (that is, the provisioning of food to wild animals by humans), some of which overlap and some of which are used interchangeably. Such terms include attractant feeding, captive feeding, diversionary feeding, necro feeding, opportunistic feeding, research feeding, supplementary feeding, tourism feeding, and unintentional feeding. These terms, and a description of how they are understood for the purposes of this chapter, are set out in Table 11.1. Of these terms, seven (attractant, captive, diversionary, necro, research, supplementary and tourism) can be considered forms of conservation feeding, while two (opportunistic and unintentional) cannot.

Conservation feeding is an umbrella term for a broad range of wildlife management activities involving the provision of food to animals. The drivers of conservation feeding are multifarious (see Table 11.1) and, likewise, its consequences are many and varied. These consequences are not always beneficial to humans, other animals, and/or wider ecosystems – sometimes intentionally, sometimes unintentionally. This chapter provides insights into different types of wildlife and conservation feeding, illustrated through a case study of feeding red kites

Table 11.1 Types of wildlife feeding. Those marked with an asterisk can also be considered conservation feeding.

Feeding type	Description
Attractant feeding*	Provisioning of food in the form of bait in order to lure wild animals to camera or containment traps for management and/or monitoring purposes.
Captive feeding*	Provisioning of food to wild animals in captivity. This feeding is usually the sole or main source of food for the animals in question and should therefore, ideally, meet all of the animals' physical, nutritional and behavioural needs with respect to diet and feeding. The animals concerned may be in captivity for a number of reasons including, but not limited to, breeding programmes, translocation or release programmes, rehabilitation, or conservation promotion via public entertainment/education.
Diversionary feeding*	Provisioning of food to wildlife to direct animals away from one area or behaviour (usually of potential or actual human–wildlife conflict) towards another (usually with lower potential for human–wildlife conflict). Diversionary feeding can include, but is by no means limited to, aversive conditioning, and is a means of mitigating human–wildlife conflict so as to avoid lethal management methods (Kubasiewicz et al. 2016).
Necro feeding*	Provisioning food in the form of poisoned bait as a form of lethal control of target species.
Opportunistic feeding	Provisioning of food to wildlife in public or private spaces to facilitate close interaction with wild animals (Dubois and Fraser 2013).
Research feeding*	Provisioning of food to wildlife for a range of reasons including: 1. To tame or habituate animals to facilitate close observation and study (Dubois and Fraser 2013). 2. To investigate home range size, survival, growth rates, behaviour, reproduction and distribution by removing or mitigating the effects of food as a limiting factor (Dubois and Fraser 2013). 3. To understand the effects of other types of intentional feeding (Dubois and Fraser 2013).
Supplementary feeding*	Provisioning of food to wildlife to complement their foraging or hunting behaviour to improve survival or reproduction and/or to anchor them to a site (Dubois and Fraser 2013; Kubasiewicz et al. 2016).
Tourism feeding*	Provisioning of food to wildlife to make 'animals predictably and reliably viewable' in order to support wildlife tourism ('non-consumptive interactions with wild animals'; Dubois and Fraser 2013). It could be argued that tourism feeding should not be included as a type of conservation feeding yet is included here since wildlife tourism is often branded as 'ecotourism' ('responsible travel to natural areas that conserves the environment, sustains the well-being of the local people, and involves interpretation and education'; TIES 2015). Thus ecotourism and wildlife tourism incorporate an element of conservation. In addition, tourism feeding can be deployed by rangers, for example in nature reserves, in an attempt to dissuade visitors from engaging in unauthorised feeding (Dubois and Fraser 2013).

Table 11.1 (continued)

Feeding type	Description
Unintentional feeding	Behaviour that creates a food source for wildlife even if the behaviour did not deliberately seek to do so, for example feeding companion animals, planting certain vegetation, discarding litter/managing waste in a way that makes it available to wildlife and so on. This feeding may be accidental, involuntary, unplanned, or even unconscious.

Source: compiled by the author from the works cited.

(*Milvus milvus*) in Britain as part of a kite reintroduction programme. The chapter is based on information collected from 2021 to 2023 during visits to five red kite feeding centres (two in Scotland and three in Wales) and semi-structured interviews with people involved in red kite feeding or wildlife policy and legislation. Participants have been given pseudonyms to preserve their anonymity, and place names and red kite release sites have been redacted for the same reason.

Kite time: feeding kites in Britain

In telling the story of red kites, conservation organisations often focus on human–kite relations in the Middle Ages, when kites were both abundant and appreciated for their presence in cities where their role as scavengers made them ideally suited to clearing up human refuse, thereby keeping the streets clean (Williamson 2013, 60). Indeed, their provision of an ecosystem service, long before such terms were thought of, saw them protected by bylaws that prohibited their killing (Williamson 2013, 60). The history and governance of kite protection and persecution is also discussed in Chapter 4 of this volume.

Human appreciation of wildlife species that are abundant is slightly, although not entirely, unusual. In general, people regard the common and familiar, if not with contempt then at least without concern, while they value the rare. This appreciation of the rare is typical within conservation, often contributing to species being endowed with charisma (Lorimer 2007), which in turn leads to those species receiving additional conservation attention to avert their local or global extinction.

Human appreciation of or contempt for wildlife species is not permanent, however, and can vary over time. Such shifts in attitudes, including in relation to the abundant and the rare, are starkly evident with respect to red kites in Britain. As outlined in Chapter 4, attitudes to red kites changed dramatically with the introduction of the Preservation of Grain Act in 1532. This act classified kites as vermin, meaning that rather than being valued for their role as scavengers, kites were seen as a pest. Financial incentives were offered for their killing and there was apparently little to no concern for the fate of this common bird.

This change in attitude towards, and attendant change in treatment of, the kite naturally brought about a substantial reduction in their numbers. Nonetheless, persecution of kites continued throughout the sixteenth, seventeenth and eighteenth centuries; in particular gamekeepers viewed kites as a threat to game birds, and large numbers of kites were killed on shooting estates. As a result, by the nineteenth century kite numbers in Britain were very low, and yet they were not afforded protection. Indeed, their scarcity made them an attractive target for egg collectors and taxidermists, thereby compounding the threat to their survival.

Eventually, in the early twentieth century, in line with conservation's convention of valuing the rare, low kite numbers precipitated another shift in attitudes towards them and, in 1903, a conservation programme was established to protect them (British Bird of Prey Centre 2023). By this stage the population was so low that without intervention the kite faced a prolonged, and even uncertain, recovery. In the 1990s conservationists therefore decided to intervene, translocating red kites from Germany, Spain and Sweden to England and Scotland. This intervention instigated new human–kite relations, which continue to evolve and which often centre on human feeding of kites. This feeding commenced with the captive feeding of kites prior to their release. It then shifted to supplementary and unintentional feeding in the short-term after the kites' release, and then became tourism and unintentional feeding – which continue in the longer term alongside opportunistic feeding. These forms of feeding are all discussed in the following sections, drawing on empirical research data.

Captive feeding and supplementary feeding

As part of the programme of reintroducing red kites to Britain, red kite chicks were captured in Germany, Spain and Sweden and transported to England and Scotland. Once there, they were held in captivity until they were ready for release: during this time humans were solely responsible for providing the kites with food (captive feeding). What the kites were fed, and how they were fed, needed to meet their behavioural, nutritional and physical needs and also to prepare them for life in the wild, including the ability to find their own food, preferably from non-anthropogenic sources. This meant 'trying to produce exactly what the wild [red kite] adults would bring in' while also trying to ensure that the food did not become 'associated with people' (Fred, conservation consultant). This latter point was particularly important since 'if you went out and fed them you could get them really tame, you'd be able to hold it [the food] and the kites would take it out of your hand' (Fred), something that is undesirable in a bird that is going to be released into the wild. The potential for habituation to humans is exacerbated by the red kites' social, gregarious nature, and their long association with and lack of fear of humans: kites have 'always followed people, since the first hunter gatherers, coming and picking up on the food that we leave behind, whether wittingly or unwittingly', they're 'a generalist species that's not really afraid of people' (George, conservation practitioner). As George is suggesting, kites have a long-standing synanthropic, commensal relationship with humans, whereby they live in close proximity to people and exploit the food resources that an anthropogenic environment provides. The current manifestation of this relationship will be discussed in the following sections.

While this human–kite relationship stretches back a long way (see also Chapter 4), it was broken in England and Scotland once kites disappeared from the landscape. As a result, a considerable amount of apprehension surrounded the reintroduction of kites, both regarding how kites would interact with people, and how people would respond to the kites. The releases were therefore very discreet: in some cases even those on neighbouring land were unaware of the release until the birds appeared in the sky and, initially, 'didn't know what they were' (Sophia, conservation practitioner). Clearly it is not possible to conceal a large and distinctive raptor once they are at liberty, nor is it possible to control their movement, although supplementary feeding of kites was carried out once they were released, in part as an attempt to anchor them to the release site and 'keep them in the area' (Sophia). This anchoring was intended to keep kites within what had been deemed favourable release sites and also to avoid potential human–kite conflict if kites left release sites and ventured into areas where they were less welcome, thus also making it a form of diversionary feeding. In some cases, the supplementary feeding was also intended to support the kites until they became self-sufficient. Indeed, Sophia suggested that supplementary feeding conducted at kite feeding stations enabled the kites to flourish, particularly by contrast with other sites where it was not carried out:

> We're the best breeding area in [place name] because I think I feed regular. I mean [name of a release site] started kites and didn't feed them and they didn't do very well. And up near [name of another release site], theirs didn't do very well, I don't think they were allowed to feed them. (Sophia, conservation practitioner)

This support of kites to enable them to thrive in their new habitats while they become accustomed to finding their own food is a major motivation for supplementary feeding, as is the hope that it will ensure reproductive success. George noted, however, that kites were able to adapt to fending for themselves very quickly: 'we used to feed the red kite but they were pretty good at finding their own food after about two months, so we just gave up'. Supplementary feeding did not always stop once it was no longer necessary for the kites' survival, however, and in these cases what began as supplementary feeding has morphed into tourism feeding for its entertainment value; animal feeding is often conducted for human entertainment as discussed in Chapter 3 in relation to flea circuses, Chapter 8 in relation to wildlife parks, and Chapter 9 in relation to zoos.

Tourism feeding and unintentional feeding

Six red kite feeding stations are now established across Britain: three in Scotland (Argaty Kites in Perthshire, the Galloway Kite Trail in Dumfries and Galloway, and Tollie Kites in the Highlands) and three in Wales (Gigrin Farm in Powys, Red Kites Wales in Carmarthenshire, and Bwylch Nant yr Arian in Ceredigion). Conservationists at these sites variously argue that the feeding they do still constitutes supplementary feeding since it supplements the food that kites would otherwise get, while also acknowledging that the feeding has now transitioned, at least partially, if not fully, into tourism feeding.

One of the elements that makes red kite feeding stations wildlife tourism is that all the sites advertise a regular feeding time, which makes the kites 'predictably and reliably viewable' (Dubois and Fraser 2013), thus maximising its entertainment value. Interestingly, even though Britain operates a daylight savings system, advancing the clocks by an hour in summer and reverting to standard time in winter, the kite feeding stations adjust their feeding times accordingly so that in real terms, and from the kites' perspective, they are fed at the same time every day, because 'the kites don't know we change the clocks' (Gigrin Farm 2023).

In most conservation feeding practices, the establishment of a routine is generally avoided; routine is, however, beneficial in tourism feeding, which depends on animals being *predictably* viewable since it is about 'bringing that experience [of seeing kites] to people' (Archie, conservation practitioner). Sophia reinforced this when she said that 'during lockdown we fed every day anyway. I mean, if you want to keep them [kites], you got to feed them regardless'. The lockdown Sophia is referring to was imposed by the British government to restrict people's movement and the operation of some businesses and other organisations as a public health measure during the COVID-19 pandemic in 2020. Sophia's point is that, despite there not being any visitors to witness the feeding, the kite feeding station continued to feed the kites, perhaps arguably for the benefit of the kites but also, as Sophia emphasises, so that the kites would still be there when visitors could return. Kites were, however, affected by the wider ramifications of lockdowns, as will be discussed later.

Kites are now flourishing in the areas around feeding stations, and the species is no longer dependent on the food provided for its continued survival. The feeding does, however, attract hundreds of kites, which creates an incredible spectacle for visitors as the kites swoop down to snatch food that has been scattered on the ground or on a sky table, which is essentially a bird table for raptors. Because the sight of hundreds of kites feeding attracts visitors, red kite feeding stations provide an excellent opportunity for conservationists to engage with visitors, and for visitors to engage with nature and its conservation. All the red kite feeding stations tell the conservation story of the red kite: its near extirpation from Wales, its extinction in England and Scotland, and its subsequent reintroduction. This allows them to talk about raptor persecution and conservation (and indeed species and habitat loss and restoration) more broadly, as explained by Ella:

> We feed at [red kite feeding location] to help the birds, particularly in the winter months, but the main reason is for educational purposes, to let people see the kites and other species up close, to raise awareness of persecution which caused the kites to nearly go extinct in the late 1900s and is still an issue in many areas today despite the fact that kites are doing well in the UK in general. (Ella, conservation practitioner)

A similar point was made by James:

> We don't need to feed the kites here, they're doing very well in other areas where there's no kite feeding going on, we look at it as an education tool to talk about the story of how we nearly lost them and the successful conservation project to keep them and how they've flourished since then. (James, conservation practitioner)

This education and awareness element of kite feeding stations justifies their classification as a form of conservation feeding, and all the sites stressed the fact that the feeding they did complemented the kites' natural foraging rather than replacing it. For example, Ella said 'we only feed a small amount each feeding day and the kites in no way rely on this, it is more like an extra snack to them'. This was echoed by other conservation practitioners, who emphasised that the kites needed to find other food: 'we see our kite feeding here as supplementary … they can come in and feed, just for a very short time each day, and get a bit of a top up but it's hopefully not detracting from what they're doing naturally, which is scavenging around the area' (James) and 'I don't overfeed, they've got to hunt for themselves' (Sophia).

The idea of 'overfeeding' was raised repeatedly, with feeding stations insisting that they fed appropriate quantities of food and being keen to distance themselves from other sites which might be doing things differently, for example, 'you do sometimes get tarred with the same brush of "these people must be shovelling out bucket loads of food for the kites every day" and of course we're not' (George). Indeed, George suggested that 'it's not so much about whether or not people do it [feed kites] … it's more about what you feed them and trying to make sure that it's not too much'. This question of how much to feed is related to the broader question of what level feeding can occur at and still be considered supplementary: 'we call it supplementary feeding and we're always keen to stress the point that it is just topping up what they find in the wild' (George). Nonetheless, in providing food for kites, feeding stations are providing a concentrated food source that, as well as being used by the kites, is also exploited by other species. Again, conservationists suggested that this was influenced by the amount of food provided. For example, James stated that 'there are other sites that will put out a lot of food … they do get more kites, they'll certainly get more buzzards at those other sites, and ravens, corvids' (see Chapter 5 for a discussion of food-mediated relationships with corvids). Meanwhile, George suggested that introducing large quantities of food into the environment could have an impact on the wider ecosystem:

> There are some places where they are feeding a lot more food than we do here and there are goods and bads about that ... at a local level that might be possibly, although I don't think it's ever really been tested, could possibly be a bit damaging to biodiversity in terms of prioritising one species over others, so more competition with other raptors for natural supplies of food, maybe impacts on whatever the prey species are as well, so that could be damaging at a local level. (George, conservation practitioner)

At an even broader level, however, both George and James pointed out that there could be an argument for prioritising the kite over other species given that, although it is now thriving in Britain, it is declining in other parts of Europe, a view supported by data from Birdlife International (2023). George went on to say that 'kites are a very important species internationally, because we've got about a fifth of the world's kite population in the UK, so what's harmful at that level might be different at an international level'. This view was shared by James, who also emphasised the importance of taking local and global perspectives in conservation:

> I know that the red kites are now extinct in North Africa, I think they were going down in Europe because the farming practices haven't been helping in some areas, so we're bucking the trend in that we're going up as opposed to the other places going down in Western Europe. So, I know that there's about 50,000 pairs they think in the world, so it's still relatively a rare bird, even though we've seen them becoming common in some areas in the UK, so we still need to watch what's going on, monitoring other areas. (James, conservation practitioner)

From this point of view, feeding kites in Britain takes on an additional conservation element in that supporting kites in Britain (and perhaps improving their breeding success) lends support to the global kite population.

On the other hand, kite feeding was also discussed as being potentially harmful, both to individuals and at a population level. Such negative consequences can be immediate if birds suffer misadventure as a result of visiting feeding stations:

> Unfortunately, a bird drowned in the lake here, basically when we got the kite out, it was a very windy day, and basically it stood there, filling its crop, it tried to fly across the lake and a gust of wind pushed it into the lake and it couldn't seem to get up and basically just the weight of the food in the crop weighed it down so much that it just couldn't get out and it unfortunately drowned. (James, conservation practitioner)

Oscar, a policy expert, suggested that negative consequences of supplementary feeding could also be much longer term and, while the detriments he mentions could affect individual birds, they also have implications for the species as a whole:

> What happens when that hunting strategy [visiting feeding stations] ... let's put it that way, what happens – when that source stops – to the hunting strategy that those birds or animals have developed? And is there a dependency in here which actually is not good? So are we increasing clumped approaches to that hunting strategy where actually having a much more dispersed hunting strategy might be more useful? (Oscar, policy expert)

Alongside an arguable dependency on feeding stations, James highlighted the threats inherent in a dependency on other anthropogenic food sources:

> During COVID, the birds had a really bad time because all the traffic stopped, people stopped moving around, so roadkill, that was all off the menu. We had that really dry warm spring, earthworms were off the menu. And lots of the birds were being handed in to the vets because they were so malnourished. So that's how they are so tied in with humans and human ways. (James, conservation practitioner)

This entanglement of kite feeding with humans and human ways can also be seen in other unintentional feeding and in opportunistic feeding, as will be discussed in the following section.

Also linked to discussion of the impact of feeding on individual birds and wider ecosystems was the question of whether sites changed the amount of food they provided depending on the season. Some sites did not change the amount they fed so as not to give the kites an advantage compared with other species:

> Our thing really is just trying to make sure that we're always putting out roughly the same amount of food and not putting out more or less even if weather conditions are harsher, some of these bad winters we get now the temptation will always be there to put out more meat to keep them going but actually that is nature and you want to make sure that you're not overfeeding them because everything's struggling in those conditions and you can't prioritise the kites I suppose over anything else. (George, conservation practitioner)

Meanwhile, other sites did change the amount they fed, to compensate the kites for seasonal variations in availability of other food: 'In the wintertime when it's heavy snow and ice I give them more feeding than what I do in the summer' (Sophia). Even sites that did not increase the amount they fed during winter months did acknowledge that the food was more beneficial to kites during the winter, and that more kites took advantage of it: 'In the winter when there's less daylight hours for feeding for the birds and they need a little bit more food to probably get them through the cold nights – colder, longer nights – we seem to get more birds coming in for this feeding, so we must be helping birds in some way' (James). This perspective again supports the argument that this feeding has a conservation element to it as well as being a form of tourism.

Regardless of whether kite-feeding stations considered their activity as supplementary feeding, tourism feeding or a mixture of both, they pointed out that their provisioning was no different from any other anthropogenic food source that kites might exploit: 'We try and stress that it's not about whether or not to feed the kites because one way or another, if they're not here, they'll be somewhere where people are supplying carrion, busy roads, wind farms, shooting estates' (George). Others drew attention to the fact that kites exploiting anthropogenic food sources is not a modern phenomenon and that:

> Kites and vultures … were feeding with Neanderthals when they were killing mammoths … so when you throw out feed for red kites, to me it's no different to a Neanderthal in a group round a massive mammoth kill, eating as much as they can, stashing as much as they can, and the rest being eaten by all the carnivores. (Fred, conservation consultant)

Indeed, intentional or unintentional feeding of kites by humans is now such an ingrained part of the kites' own feeding behaviour that Sophia conflated the two: 'in the summer the kites are following tractors for harvest, hay, silage harvest, so they don't come in here from four, half past four, five o'clock, they're getting *natural* kill' (emphasis added). This human–kite entanglement will be returned to in the discussion of the kite's future in Britain.

Opportunistic feeding and unintentional feeding

Arguments for tourism feeding being a form of conservation feeding (as well as, or despite, having an entertainment value) are that it is controlled (in that it is conducted by, or under the supervision of, conservation professionals, and food is provided in appropriate quantities and is of an appropriate nature), and that it dissuades people from engaging in opportunistic feeding by providing an alternative opportunity to engage with wildlife (Dubois and Fraser 2013). In the case of red kites this latter point bears consideration, with conservation practitioners highlighting that opportunistic feeding of kites did take place, in some cases very close to feeding stations:

> We sometimes see up to twenty-five kites in the village coming down and feeding off where they park the car normally or in the back garden, and there's another one just across the road from me and that lady, she probably gets ten or a dozen kites coming in. And you see that in quite a few other areas, once one or two kites find it and if you start to regularly feed them, then they'll keep coming back and then others will catch on the idea and then you get more and more birds coming in, so there's definitely some of this sort of feeding going on in other places and other villages. (James, conservation practitioner)

Clearly tourism feeding is not entirely eliminating opportunistic feeding of red kites – indeed it is possible to extrapolate that the former may even be driving the latter, if people are tempted to replicate their experience at a feeding station at their own home. The desire for repeated, close interaction with kites may be exacerbated by the fact that kites are distinctive, charismatic birds and are not particularly anthropophobic. In addition, given that kites were reintroduced to areas from which they had been extirpated, their renewed presence is (at least initially) novel and exciting to some people (Orros and Fellowes 2014).

Despite its appeal from the human perspective, conservationists questioned the benefits to kites of this opportunistic feeding, and also of unintentional feeding: 'There are some parts of the UK where the kites, their chicks are being hatched with birth defects because they've been scavenging or been fed processed meat' (George). Unintentional feeding can also be negative for the people involved: given their lack of fear of people and their generalist diet, kites have been reported as 'stealing food from children's hands' (James), leading to concerns regarding human–kite coexistence. Given the expanding kite population, this coexistence is a major question in relation to kite conservation in Britain.

A kite future?

As with all human–wildlife coexistence, human–kite coexistence can be harmonious but can also result in conflict. The case of the red kite in Britain does however represent something of a special case because kites are a reintroduced species. Coexistence with reintroduced species has been identified as being of a different quality from coexistence with an extant species, largely because people in the area to which the species is reintroduced are unlikely to have experience of living with it (Auster et al. 2022). Lack of experience with, or loss of tolerance for, extirpated species can complicate attempts at *renewed* coexistence. Auster et al. (2022) posit renewed coexistence in these circumstances as requiring people to *re*learn how to live with the species in question. In this respect, renewed coexistence can be related to attitudes to rewilding, where people are required to develop a renewed, or even radical, tolerance for the agency of other species and to adopt an enlightened attitude to wildlife (Arts et al. 2016; Campbell 2006).

A significant aspect of renewed human–kite coexistence will relate to how humans feed kites, whether intentionally or unintentionally. Since their reintroduction, kite and human lives have rapidly become entangled, and much of this relates to the fact that 'people provide the bulk of their [kites'] food one way or another, be it from deliberate feeding or roadkill' (George); the ways people feed kites are complex, with supplementary feeding, tourism feeding, diversionary feeding, opportunistic feeding and unintentional feeding all overlapping and evolving as kites re-establish themselves in the landscape. Human–kite coexistence, including human feeding of kites, will continue to evolve as the kite population continues to expand. In living memory, kites in Britain have shifted from being almost extinct, through being rare and novel, to becoming (at least locally) ubiquitous, and demanding a place in our consciousness and the landscape. This raises potential challenges in light of the fact that we have:

… an expanding kite population, but you've also got a massively inflated human population as well so there's not much space for nature to be … in Britain I think we're not particularly used to having to live alongside wildlife … and kites push that a little bit and test people's patience in some of these places. (George, conservation practitioner)

Mutual flourishing of humans and kites in Britain and the establishment of harmonious, renewed human–kite coexistence thus depends on these two expanding populations relearning to live together, something that will require greater human tolerance for kites and their agency, particularly in relation to their opportunistic feeding behaviour.

References

Arts, Koen, Anke Fischer and René van der Wal. 2016. 'Boundaries of the wolf and the wild: A conceptual examination of the relationship between rewilding and animal reintroduction', *Restoration Ecology* 24: 27–34.

Auster, Roger E., Stewart W. Barr and Richard E. Brazier. 2022. 'Renewed coexistence: Learning from steering group stakeholders on a beaver reintroduction project in England', *European Journal of Wildlife Research* 68 (1): 1–22.

Birdlife International. 2023. 'Red kite (*Milvus milvus*)'. Accessed 11 June 2023. https://datazone.birdlife.org/species/factsheet/red-kite-milvus-milvus.

British Bird of Prey Centre. 2023. 'Red kite'. Accessed 11 June 2023. https://www.britishbirdofpreycentre.co.uk/conservation-projects/red-kite/.

Campbell, Timothy. 2006. '"Bios", immunity, life: The thought of Roberto Esposito', *Diacritics* 36 (2): 2–22.

Dubois, Sara and David Fraser. 2013. 'A framework to evaluate wildlife feeding in research, wildlife management, tourism and recreation', *Animals* 3 (4): 978–94.

Gigrin Farm. 2023. 'Opening times'. Accessed 11 June 2023. https://gigrin.co.uk/opening-times/.

Kubasiewicz, Laura M., Nils Bunnefeld, Ayesha I. T. Tulloch, Chris P. Quine and Kirsty Park. 2016. 'Diversionary feeding: An effective management strategy for conservation conflict?', *Biodiversity and Conservation* 25: 1–22.

Lorimer, Jamie. 2007. 'Nonhuman charisma', *Environment and Planning D: Society and Space* 25 (5): 911–32.

Orros, Melanie E. and Mark D. E. Fellowes. 2014. 'Supplementary feeding of the reintroduced Red Kite *Milvus milvus* in UK gardens', *Bird Study* 61 (2): 260–3.

The International Ecotourism Society. 2015. 'TIES announces ecotourism principles revision'. Accessed 25 November 2024. https://ecotourism.org/news/ties-announces-ecotourism-principles-revision/.

Williamson, Tom. 2013. *An Environmental History of Wildlife in England 1650–1950*. London: Bloomsbury.

Conclusion. In conversation: non-utilitarian feeding, interdisciplinarity and the future of feeding research

Herre de Bondt, Hannah Britton, David Cooper, Gaia Mortier, Hannah C. Mortimer, Felix Sadebeck, Virginia Thomas and Juliette Waterman

This volume has brought together authors and themes from a variety of disciplines, all under the umbrella of the Wellcome Trust project 'From "Feed the Birds" to "Do Not Feed the Animals"'. In planning the volume and considering how to summarise our arguments, there were few solutions that satisfied the breadth and depth of questions that arise from multi-disciplinary research conducted by early career researchers (ECRs). Inspired by the conversation pieces published by the *American Historical Review* (Hoffman et al. 2008), we decided to frame this conclusion as a discussion piece between several of the authors and editorial team, with additional analysis around the core topics that arose. The conversation was conducted via video call in November 2023, with the transcript recorded and edited for clarity.

We began this volume by raising the important themes of non-utilitarian feeding (with a wealth of associated considerations, interactions and implications), interdisciplinarity, and future avenues for research and discussion: we returned to these as the basis for our discussion. As well as drawing together some of the crossovers between our chapters, we were also keen to explore the implications of the project findings for future feeding practices.

Joining Herre de Bondt, human–animal ethnographer and co-editor of this volume, were several of the volume's contributors:

Hannah Britton, zooarchaeologist; David Cooper, natural scientist; Gaia Mortier, entomologist; Hannah Mortimer, human–animal sociologist; Felix Sadebeck, zooarchaeologist; Virginia Thomas, sociologist; and Juliette Waterman, archaeological scientist.

Herre
While preparing for this discussion, I went through all of our chapters trying to draw out and synthesise the main arguments we were each trying to make. It was quite difficult, in a good way, because we are using this book to generate a lot of new questions. Many authors used their chapters to ask questions such as: What if we look at feeding from an affective perspective? How does our analysis of practices, sources, and phenomena change if we bring feeding to the foreground? How can we rethink feeding as an action of ecological intervention rather than a dialectic relationship between human and non-human animal? Generating such questions opens up further lines of enquiry for discussion and future research, which is very fruitful in itself.

But to move it towards more of a concrete topic, I think we should talk about how the project is based on the concept of 'non-utilitarian feeding'. Hannah Mortimer's chapter [7] reminded me that Virginia Thomas and Angela Cassidy, one of the co-investigators on our project, define this as 'human feeding of other animals that is not part of a direct transaction or a means to an end' [Thomas and Cassidy 2022]. This definition, as well as many of our chapters, demonstrates that it stands in direct opposition to utilitarian feeding characterised by serving a clear transactional purpose, as we see in systems of production. So what a lot of us wrote about are issues of control. What is feeding doing? How is feeding affecting animal behaviour, animal morphologies, and in what ways is feeding – either directly or indirectly – influencing animal lives?

Gaia wrote one line [Chapter 3] that particularly touches on this question of how feeding controls animal populations. She wrote: 'If humans pick and choose what nature is allowed to continue, the world will contain nothing more than "functional" biodiversity.' And so I wanted to just throw that out there to get the discussion started: what are your thoughts on the terms utilitarian and non-utilitarian and on Gaia's quote?

Felix
This '"functional" biodiversity' is a very interesting phrase. One thing that really comes across in a lot of these chapters is a discussion of how we see different species. At what point are they considered vermin

and when are they animals that we appreciate and that we like to have around? Such perspectives are changing all the time, just like Juliette's chapter [4] shows us. She writes about how we have come full circle in Reading, where red kites were initially seen as a pest but after rigorous persecution people started missing them again. However, now that they have been reintroduced, they are slowly swinging back to being seen as a pest. Based on that, we can expect that, whatever we do now, our actions might lead to irreversible changes to biodiversity. And if people in the future think 'Oh, actually, I would like to have more pigeons around' – maybe it's too late. Pigeons might not be the best example since they're so ubiquitous, but you get the point. And Riley's chapter [5] shows how difficult it is to come back from such a pathway. There are still areas in Britain where you essentially don't have any ravens because they were excessively hunted in the past, and even though they're not hunted any more – or at least not officially – it just takes a while until they come back, and maybe some of them won't come back properly.

Hannah Britton

When Naomi Sykes, our principal investigator, was pitching this project, she mentioned the ways that animals are perceived based on their rarity. When a lot of species were introduced into Britain, they were considered like gods, and it was a spread of culture, and then they became more common, and then they became pests, or they became 'normal animals' like the chicken and other species [Sykes 2012]. And I think, for a lot of species, how common they are and how rare they are has a really big impact on their social value. And what we see is that this often comes down to conflict for food or feeding practices.

Felix

Which is so ridiculous, isn't it? Because as you pointed out, the number of animals of a certain species that we encounter changes how we perceive them. But then – as the example of red kites shows us – you start feeding them because you think they're rare and impressive, but as their population increases they're suddenly not that remarkable anymore because you fed them, and then you stop feeding them and start persecuting them. Animals' images and populations are so human-driven, and we don't even notice how we are actively working and changing future perspectives of these animals through feeding them.

Virginia

Can I say something about '"functional" biodiversity', because this comes up in rewilding and reintroductions a lot? I think functional biodiversity reflects a very utilitarian view of nature. If animals are extinct, where there are extant animals that can and do play a similar role, and fill a similar niche, we might reconsider some of our conceptions of nativeness. Similarly, with reintroductions, we might consider not necessarily reintroducing species for an instrumental reason, but for an intrinsic reason, for their own sake. People used to ask Gerald Durrell what the point of certain species involved in reintroductions were, for example: 'What's the point of red kites? Why do we need red kites back in Britain?' His answer was: 'What's the point of you?' Does it matter what the point of the red kite is? Perhaps ravens or buzzards are already fulfilling the role that the red kite plays in the ecosystem, but does that matter? Maybe the red kites should be back in Britain for their own sake and for their intrinsic value rather than any instrumental value.

Gaia

A couple of weeks ago Juliette and I were talking with Dr Ollie Douglas, the curator of the Museum of English Rural Life, and he gave us an interesting perspective by comparing extinction to trades. So for example, if a very old trade, like shoemaking, is going extinct in certain areas of the world, we tend to see it as a really negative thing. But the trade could be picked up, continued, or developed somewhere else again. So the trade itself is not necessarily going extinct. Coming back to animal extinctions, red kites maybe went extinct in England, but were very much still alive in other areas of the world. Obviously that was a significant development to many people in England, because red kites used to be there and then they weren't. It is this local view of extinction that made people consider it as an issue, while a global view on extinction paints an entirely different picture.

Hannah Britton

I think it's also interesting about the concept of whether they want to be here or not. If one were to introduce a bird, say, a golden eagle or a white-tailed eagle, in a specific area because humans determine that it belongs there, it might just decide to fly away. It could simply say 'no, I don't want to be here', and defy these human ideals and ideas. And I think that's quite an interesting sort of thing of how – even though rewilding is sometimes trying not to manage it that much, I don't know if Virginia has opinions on this – sometimes we are managing it. And they're animals:

sometimes they just want to do their own thing – but obviously if you're a mammal, you can't really run away as much as maybe a bird can fly away.

Herre
I wanted to also bring feeding back in a little bit, because feeding often controls animal populations and movement. In Chapter 11 Virginia gives an outline of nine different kinds of wildlife feeding – which can be attractant, captive, diversionary, necro feeding, opportunistic, research, supplementary, tourism, or unintentional. All those things are very much about controlling animal populations in terms of whether they live or die, how they live, and where they go. Even unintentional feeding implies control somehow, because the feeding is very much on human terms: we didn't intend to feed and influence these animals, but we still consider it human-instigated *feeding* rather than opportunistic eating from the animal's perspective. Anthropocentrism is so inherent in the term feeding that, even if we consider it unintentional, humans have to be the driver of changes in non-human lives.

Gaia
Speaking of intention, I think one of the most interesting recent phenomena from an entomological perspective is that people are pushing really hard for wildflower patches and planting flowers in your garden specifically to benefit certain species of bees or butterflies. And in itself, that is quite positive, but it is also perpetuating that narrative whereby we are picking species again, as only specific species would be attracted to those specific flowers. Many people are particularly targeting honey bees, which aren't necessarily the species that you get with all those flowers. So it's a very interesting dynamic now, the idea that if you are planting a patch of flowers you will attract more bees in your garden, while in reality you are attracting and feeding far more than just those bees. And yet people are doing it with that species in mind.

Felix
I think that's a super-interesting point because this is just one example of how detached this process of feeding can become from the original intent and intentions of feeding. Some people will just plant some wildflowers in their garden because they want to feel good; they might even say to themselves 'Well, this is part of my duty to fight climate change, I'm one of the good guys'. And in that process they are not even thinking about all the various animals that they are attracting and that they might be unknowingly contributing to insect species selection. There's a huge

difference between the original intention of people's feeding practices and the actual results of said actions. There certainly could be a lot more reflection on the consequences of this feeding process. On the other hand it is good to consider that it's not inherently good or bad. If someone wants to feed pigeons because it makes them feel good, I think that's a damn good reason just all by itself.

Herre
So far we talked about feeding and control: specifically the effects it's having on populations and behaviour. But Blessing's [10] and David and Andrew's [9] chapters describe how feeding influences animal morphology. Blessing writes about how metal pollutants end up being stored in foxes' bodies while David and Andrew talk about how feeding is changing the bone structure of animals within a species. Both of these are very much a consequence of feeding that a lot of us are not talking about as much. Reading their chapters made me think particularly about our ideas of what animals are.

David, you wrote about how captive populations are becoming distinct from wild populations. Reading that made me think a lot about the concept of species, and how feeding is then influencing our idea of species, where animals are drifting away from their wild counterparts, and to what extent we should rethink species in such a context.

David
I suppose a lot of what I'm talking about isn't this kind of commensal domestication, evolutionary change in the species. It's mostly about what animals eat and how it changes their morphology plastically – meaning within the lifetime of the individual. What we're seeing is that because captive animals are eating different things from their non-captive counterparts, they become bigger or the shape of their skull changes. And while you might be like 'Oh well, that's not that interesting because it's just in the individual, we're not domesticating them', it can potentially still create a dependence where those individuals are less fit for potential reintroductions.

I think something like this came up in some of our original proposals about feeding the swans and ducks. When people were told 'don't feed the birds', those swans and ducks died because before the feeding-ban they were held at a population that's higher than would be natural in those environments without human feeding [Horton 2018]. So those animals are less suited to a natural situation as a result of populations increasing, but populations might similarly struggle to survive because their morphology changes as the musculature of their jaws changes, as a result of what we're feeding animals.

Felix
It's also an interesting marker for archaeologists who want to understand the past relationships between humans and animals. You can pick up specimens that make you think 'these two skulls are of different species because their morphology is different', where actually it could just be that they're the same species that has been fed completely different foods. Bone morphology contains interesting information all by itself that might have gone unnoticed in the past.

Hannah Britton
My research is almost a reversal of that. Where you're talking about accidental morphological change as a result of feeding, falconry deliberately feeds birds the right things to ensure that they retain specific morphological elements. With falcons, they have a tomial tooth in the beak which is used for ripping flesh. If falcons are not eating proper meat, the tomial tooth can overgrow and can split and cause a lot of infection and disease and the falconer would have to manually shave them down. While this does on rare occasions happen in the wild, it happens commonly with captive birds because their food does not have the proper consistency. And so you have to feed your bird to ensure that they *don't* have skeletal and soft tissue changes happen to them.

Hannah Mortimer
We've not actually talked about livestock farming at all yet, but this discussion reminds me of what some of my interlocutors have been saying about how grass-based or pasture-based farming systems are more natural than others. So they're trying to feed the livestock what those animals would be eating if they were actually living in the wild, if they weren't domesticated. I think that touches upon a very core question of what is the right way to feed animals.

Felix
It also raises the question of what would have been the natural diet of the undomestic form? A classical Eurasian aurochs wouldn't eat dried hay in the winter for months – so farmers feeding their cattle grass throughout the winter is upholding an artificial perspective on what is perceived as being a natural diet.

Hannah Mortimer

A lot of farmers I've spoken to think it's more natural to have ruminants grazing during the winter as well. One of the reasons why they house them in barns in the winter (where they get fed preserved forage like grain, hay and silage) is because otherwise their hooves would 'poach the sward' (that is, damage the grass and underlying soil) in wet weather. This is especially a problem in Devon where the soil contains a lot of clay and therefore a lot of water, as opposed to other places which have free draining soils – I do know of farmers in other parts of the country who are outwintering their herds. Another thing is that cattle can get hoof rot and other illnesses if they're left outside in wet, muddy conditions. The farmers are caring for multiple things – animals and environment – and it's all connected.

Felix

It's funny when we compare this to what Roman literature tells about feeding cattle, because the Romans were incredibly fast to say 'actually if you want to raise more little cow babies than the mother is capable of sustaining with milk, then just grind up some basically peas, and mix them with water and give it to them – it's just as good as milk'. There's not a tiny speck of thinking about whether it's natural what the cows eat. The attitude was 'it works, you can use this as well, then you have a more or less healthy cow afterwards as well'. So what I wrote in my chapter – if the cattle is ill, you feed them basically human food. It's almost the opposite of trying to find the natural diet: it's trying to figure out what works in a very human-centric way.

Gaia

I think I remember you saying this, Hannah, correct me if I'm wrong, there was such an interesting dichotomy of people that have livestock that feed them the most natural diet, because they think that's the optimal way to do it. But then on the complete other end, you have people that use artificial food that may offer the most complete diet with the right vitamins, nutrients, et cetera, which they think is the best strategy. That shows us that there's two polar opposites. It's a bit like the question of: would a wolf be happier out in the wild where it needs to struggle to survive, or is it happier in a house where it has all the food it wants but is locked in and not in the wild?

David

But I think part of the issue with zoo feeding over the last fifty years or so is trying to make these scientifically nutritionally complete diets that are mechanically soft and useless and affect development. So I think the issue is that, maybe in an ideal world you can have the best food, but actually coming up with that is more complicated than I think most people have traditionally thought. And in a way – the lazy way – giving animals or ourselves the best thing to eat is just giving them what they've evolved with in the wild naturally because that's where they should be.

Virginia
We're in this situation where we've got the idea that 'natural' food is good and that's what animals *should* be eating. But they're in really artificial environments so it doesn't necessarily fit. Actually feeding them their natural food in a captive artificial environment isn't necessarily the best thing for their behaviour. It might be the best thing for them nutritionally and physiologically, but not behaviourally.

Felix
It's quite funny, isn't it? First we capture these animals, we domesticate some of them and turn them into a totally new creature, and then we're trying to kind of give them as-natural-as-possible food. It doesn't make too much sense when you could also just embrace the fact that you keep it in a zoo and think 'I keep it in a zoo, so I feed it zoo food', right? But it's almost as if there's a kind of guilt in there – a feeling of guilt that people try to compensate for through feeding. I think this also links with Roman cattle – if the cattle is not well and you have an empathic bond, and you feel not good because your cattle are not healthy, you would give human food to them and give it in a certain way. Because you want to release yourself of this feeling of guilt, so you try to give something human – something 'good' – to it.

Herre
What I would also like to talk about is how feeding shapes relationships between animals and humans. Juliette's [4] and Riley's [5] chapters present historical deep dives into how feeding fuels conflict or coexistence. One quote I really liked, talking about conflict for food, was Riley's quote: 'food frequently [is] … the wedge driven between our species'. That made me wonder to what extent feeding is contributing to conflict and negative experiences and connotations for both humans and animals.

Juliette
That's something that's been happening with red kites in Reading lately. While Reading's residents claim that feeding is making them feel close

to the animals, they simultaneously worry about red kites becoming a nuisance. Local media report about red kites stealing meat off barbecues [Ffrench 2020] which makes some people reduce the amount that they feed. There seems to be a growing concern that feeding might also lead to negative effects as red kites – or other species that feeding attracts such as rats – cause havoc on human residents. But the other side of this conflict is that it often results in humans taking measures against animals.

Virginia
I'm not sure if this is quite the same, but with animals that you're feeding for release you are deliberately creating, if not a negative relationship, at least a distant one. Which is quite different from the kind of intimate relationship that you might associate with feeding. So with preparing red kites for release, you're specifically trying to avoid them associating people with food. Everything changes once they're out in the wild, but before that you build cages especially so you can put an arm in to feed them without the kite seeing you, or with some kind of compartment that you can put food in while the kite's in another compartment, and then open a hatch and allow them in to the food. My participants specifically said that that's a risk with red kites because they're a social raptor compared to other raptors. They're also not very anthropophobic, and quite opportunistic, generalist birds. Because of that you could get them very tame, very easily, and you're absolutely trying to avoid that, because that's the opposite of what you want.

Hannah Britton
Related to that, there's a reason why buzzards are used in falcon displays a lot and it's because they're so tame. Even though they're not really hunters and if you want to show a hunting bird it would be better to have something like a sparrowhawk, goshawk or peregrine, because they are hunters and that is their nature for falconry. But those birds sometimes will have tantrums and refuse to listen, whereas birds like kites and buzzards are so much easier to tame and have a better relationship with humans, which is exactly why they are often used in displays. That worry about anthropogenic associations with food is the reason why medieval people would leave Harris hawks in the nest for so long before getting the birds out. They don't want the birds to create this human–food association. Instead, they want to use food as a currency for a relationship rather than as a parent providing the food, basically.

Juliette

I'm really interested in what drives certain animals to be more likely to take a risk to scavenge on anthropogenic foods. I know there's an element of scavengers versus predators, but it seems like there's maybe a little bit more to it with certain taxa as well. Certain animals are much more likely to come into close contact and be tolerant of humans and be interested in what we're putting out into the environment. Obviously for birds of prey it mostly comes down to scavengers, but I think there are other aspects of it as well.

Virginia

I think your question's a really good one Juliette, because in this project we talked a lot about why people feed animals but we haven't really talked about why animals *take* food. There's been work done on fallow deer in Ireland and it seems to be that the bold and more gregarious fallow deer are taking more food [Griffin et al. 2022; Griffin and Ciuti 2023]. But then you're almost artificially selecting for those traits: individuals who have those bold traits are benefiting more from anthropogenic feeding, and therefore more likely to reproduce successfully.

Juliette

That's really interesting as well, because it all seems to come down to personality. We tend to shy away from anthropomorphisation, but individual differences and personalities of animals is heavily underconsidered in how it relates to feeding. What you were saying there, Virginia, about the deer approaching things – I remember Riley telling me a similar thing happening with corvids [Greggor et al. 2016]. Because what started with a few bold individuals turned into entire populations that are exploiting more and more food resources, they're able to go more and more places, they're able to eat in more places, so they can breed in more and more places, and so on and so forth.

Herre

What I like about this part of the discussion is that it is bringing the animal back in a little bit, because I think feeding inherently is very much about the human, whereas eating is the animal part of it. In a lot of this discussion as well as our writing, it's difficult to keep the animal in as an individual with agency. While it is true that humans are feeding, we should not forget that there are animals that are choosing to eat. We also shouldn't forget to talk about interdisciplinarity, since this project was an interdisciplinary endeavour from its very inception. How did you experience working with other disciplines, and how does cross-discipline collaboration influence our findings?

David
You know what, I don't like the term interdisciplinarity, because I think this should rather be called a multidisciplinary project. Arguably, interdisciplinarity has not gone very well within the project, but I think that's not necessarily a bad thing. I think with interdisciplinarity, you end up watering down interesting points of view from all parties. I think this book is more multidisciplinary, and it's interesting for researchers to then look at these different perspectives in one place, to actually understand different perspectives on the same topic. I think as soon as you start trying to make this smudge-together of interdisciplinarity, you lose something in that. So yeah, I do see relevance, especially in the question of why we feed animals. But ultimately what I end up doing myself, I do sit in my own little niche and write about skulls.

Gaia
I actually think interdisciplinary does hold up when you think about it. But maybe more in the sense of the saying 'one person can't do science'. So if for example I was looking at this incredibly niche topic of parasite feeding, but even that niche topic has many different perspectives you can approach it from. I see interdisciplinarity on this project as building a 3D image with 2D perspectives from different angles. Especially because we're combining people that are really focused on the animals – so really biology, morphology, zooarchaeology, but also with the contemporary human aspect of it. It feels like our work is a lot more backed-up, rather than just one person expressing their findings and opinions from one single angle.

Herre
I've noticed a similar thing in my research on red kites. Juliette informed me about red kites, and I really needed that biological perspective in order to do anthropological research. Both in the sense of knowing what I'm talking about, and having participants have a sense that I know what I'm talking about. Maybe to me interdisciplinarity mattered in the research design step. My research questions very much depended on what I knew about red kites, of which my knowledge was very limited as an anthropologist.

Juliette
The opposite is also true, where archaeology can learn a lot from contemporary perspectives, as Virginia's chapter [11] on red kites shows.

I started with the research question that was really thinking about how people were engaging with animals in the past and what that relationship would have looked like. But obviously, as with anything archaeologically, it's very hard to go anywhere, even in a very subtle way, if you don't have a good understanding of what that looks like in the present. And there is a big danger in archaeology, and people are always very cautious especially with early archaeology to avoid comparisons with modern ethnographic cases, because in a lot of cases it can be really problematic to say 'oh you know, people today do x, y, z, and therefore when you see this evidence in the archaeological record, it means they were doing the same thing' [Gosselain 2016; Lyons and David 2019]. Rarely is it that simple, but even so, having a good understanding of what those human–animal relationships look like, and what kind of drivers, factors and impacts are involved in modern conservation and modern persecution, does really set the scene for what you're seeing in the archaeological record, for sure.

Felix

Throughout my research on Roman cattle I experienced something similar whenever I talked to Hannah Mortimer. I have no idea about modern cattle, but it's interesting to see that there is actually quite a change of perspective going on, and I would have missed that if I just looked at Roman literary sources and old cattle bones. It's one of those examples of 'absence of evidence is not evidence of absence'. Just because you don't see evidence of human–cattle bonds in the archaeological record directly, doesn't mean it's not there. Maybe you haven't looked at the right angle. What I discovered when looking at these sources for medical feeding of cattle was that actually an understanding of relationships between cattle and humans is needed in order to try to reinterpret these sources.

Hannah Britton

That really links to when I'm talking about the health of birds, particularly with veterinary feeding. Adelard of Bath's twelfth-century writings are very instructional [Burnett 1998] – but I know that, when my birds are sick, how that relationship is when you're feeding them. What it sounds like, what it smells like, what such a relationship is actually like. And when I'm reading historians writing about the past or archaeologists making interpretations, I immediately go: 'Okay, well they've missed this really obvious thing, and if you've ever spent several hours with the young bird or training a bird, you would know this'. Especially veterinary feeding is very much a sensory experience, and behavioural aspects really impact that. Unless you've actually had that physical experience, it is really hard to perceive if you're just someone in a lab or reading text about it.

Herre
I just wanted to bring in David because I remember you talking about your research on chimpanzees and potential research on differences between wild populations. It reminds me of anthropological and ethological collaborations on different populations of Barbary macaques and how humans and macaques come into conflict with one another [Majolo et al. 2013]. That research looked at the human side as well as the macaque side of increasing tourism and feeding practices to provide a more complete image. Do you think your morphological research lends itself to such interdisciplinary collaborations as well?

David
What I'm looking at is using a captive/wild situation to understand plasticity – how much something can change within a lifetime – and that's really interesting to understand wild populations living off natural diets. If you're looking at an animal that's got a continental scale range, populations living in northern regions will probably be eating something different to southern populations just because the environment is different. So the environment is driving morphological changes between them. But then it also acts as a good experiment, as Herre was saying, to look at how people are feeding different animals in the wild in different places, and how maybe there's not a genetic component to that – we're not changing animals or domesticating them, but we are creating these different morphologies. One of the ways that's expressed is with the foxes in London, as in Blessing's chapter [10], and how what we're accidentally feeding foxes in London might be kind of changing what they look like, essentially. But foxes have only come into London in the last 60 or 70 years so we just don't know whether those changes are related to founder effects, rapid genetic changes, plastic effects, environmental effects, or something entirely different.

Herre
Maybe interdisciplinarity entails interaction and collaboration across disciplinary borders while multidisciplinarity pulls together knowledge from different disciplines while staying within its own boundaries [see also Choi and Pak 2006]. Working together with other disciplines while

designing and conducting your research certainly adds a more complete picture, or a 3D picture as Gaia suggests. But at the same time David is also making a good point in the value of writing strong single-discipline pieces and presenting multiple perspectives alongside each other for other researchers to draw their own conclusions from. But what about future perspectives on feeding? This project started out with a very handy name: 'From "Feed the Birds" to "Do Not Feed the Animals"'. That title implies a bit of a shift from feeding animals as normal to the idea that maybe we shouldn't feed animals. What do you think?

Juliette
I think it was very telling that, when we organised the birds conference with the team from the British Trust for Ornithology, and we asked them the same question, most of them said that, despite the caveats for bird feeding, most of them do feed birds. I thought that was very telling.

David
I think in the future we might increasingly have to do conservation feeding. I thought it was quite interesting with the kites, because kites could be an umbrella species for habitat and for functional ecology or morphology. In conservation feeding you're not actually improving habitat, you're just feeding your umbrella species and keeping that going. And I wonder how much in the future with habitat loss and biodiversity crises, we end up feeding charismatic wildlife in our cities, rather than trying to promote the habitat and promote that side of conservation.

Gaia
I think we have to be conscious of which species are the most charismatic and, as a result, will receive the most feeding. You see this a lot in insect conservation, where 9 out of 10 papers are focused on either beetles, bees or butterflies – the 3 Bs – because that's the ones we see and that's the ones the general public recognises. Those are the ones that we accept as 'the insects'. When I do outreach for insects specifically for conservation or anything, people come up to me and they ask: 'How can we help? How can we get insect diversity?' And what I tend to say is 'get a log and throw it in your garden and forget about it' – because that, from a biodiversity standpoint, is probably the best [Goddard et al. 2013]. It just means you won't get butterflies, bees, beetles necessarily – you'll get all the small things that you don't see that are just as important. But they are obviously looking for an answer that's kind of like 'plant more flowers' or 'put a little dish of water out', because that's a much more direct form

of feeding. It's interesting that that's not the official advice from a lot of avenues that are promoting biodiversity.

Hannah Mortimer
That holistic look at feeding really overlaps with my work as well. A lot of regenerative farmers do use the word 'holistic' to describe their system, so they'll have wildflower hay meadows and herbal leys – so they're not just feeding the cattle or the sheep on that, but feeding other species, like bees and butterflies. And they talk quite a lot about how their kind of farming is 'nature friendly' or 'farming in harmony with nature' because it promotes biodiversity. The indirect feeding is just as important to them, as you're aiming to feed one species, but you end up feeding lots of others at the same time.

Herre
Kind of interesting to think about that holistic perspective, because in bird feeding I think there's a similar trend towards wildlife gardening, where you just plant a bunch of plants in your garden to create a healthy ecology. Maybe we should think of feeding as supporting an ecology, instead of direct feeding. Feeding has a connotation of direct action and reaction. Whereas there's a whole ecology – which Giovanna and I also tried to point out. There is more than just the human and the animal that is being fed. And maybe it would be good for there to be more of a shift towards nurturing ecologies rather than nurturing species.

Felix
I think it's a brilliant point. If you think about feeding, obviously we can't give a simpler answer to 'is feeding bad or good' or whatever, but to take it from this point of view that if you do feeding, or want to do feeding, or are considering doing feeding, then consider the whole bunch. Don't kind of fall back to this assumption of 'I'm just feeding the stray cats' – no, you're not. So if you do it, it's fine, but be aware of what you're doing and think about the consequences.

References

Burnett, Charles (ed./trans.). 1998. *Adelard of Bath: Conversations with his nephew*. Cambridge: Cambridge University Press.
Choi, Bernard C. K. and Anita W. P. Pak. 2006. 'Multidisciplinarity, interdisciplinarity and transdisciplinarity in health research, services, education and policy: 1. Definitions, objectives, and evidence of effectiveness', *Clinical and Investigative Medicine* 29 (6): 351–64.

Ffrench, Andrew. 2020. 'Red kites attack barbecue to steal food', *Oxford Mail*, 30 September. Accessed 12 February 2024. https://www.oxfordmail.co.uk/news/18758354.red-kites-attack-barbecues-steal-food/.

Goddard, Mark A., Andrew J. Dougill and Tim G. Benton. 2013. 'Why garden for wildlife? Social and ecological drivers, motivations and barriers for biodiversity management in residential landscapes', *Ecological Economics* 86: 258–73.

Gosselain, Olivier P. 2016. 'To hell with ethnoarchaeology!', *Archaeological Dialogues* 32 (2): 215–28.

Greggor, Alison L., Nicola S. Clayton, Antony J. C. Fulford and Alex Thornton. 2016. 'Street smart: Faster approach towards litter in urban areas by highly neophobic corvids and less fearful birds', *Animal Behaviour* 117: 123–33.

Griffin, Laura L. and Simone Ciuti. 2023. 'Should we feed wildlife? A call for further research into this recreational activity', *Conservation Science and Practice* 5 (e12958): 1–15.

Griffin, Laura L., Amy Haigh, Bawan Amin, Jordan Faull, Alison Norman and Simone Ciuti. 2022. 'Artificial selection in human–wildlife feeding interactions', *Journal of Animal Ecology* 91 (9): 1892–905.

Hoffman, Richard C., Nancy Langston, James C. McCann, Peter C. Perdue and Lise Sedrez. 2008. 'AHR conversation: Environmental historians and environmental crisis', *American Historical Review* 113 (5): 1431–65.

Horton, Helena. 2018. 'Birds are starving because people are told not to feed them bread, warns Queen's swan guard', *The Telegraph*, 14 November. Accessed 12 February 2024. https://www.telegraph.co.uk/news/2018/11/14/campaign-against-feeding-ducks-bread-starving-birds-warns-queens/.

Lyons, Diane and Nicholas David. 2019. 'To hell with ethnoarchaeology… and back!', *Journal of Archaeological, Ethnographic and Experimental Studies* 11 (2): 99–133.

Majolo, Bonaventura, Els van Lavieren, Laëtitia Maréchal, Ann MacLarnon, Garry Marvin, Mohamed Qarro and Stuart Semple. 2013. 'The singular case of the Barbary macaque'. In *The Macaque Connection: Cooperation and conflict between humans and macaques*, edited by Sindhu Radhakrishna, Michael A. Huffman and Anindya Sinha, 167–83. New York: Springer.

Sykes, Naomi. 2012. 'A social perspective on the introduction of exotic animals: The case of the chicken', *World Archaeology* 44 (1): 158–69.

Thomas, Virginia and Angela Cassidy. 2022. 'Practicing engaged research through pandemic times: Do not feed the animals?', *Journal of Science Communication* 21 (2, A05): 1–20.

Index

References to figures are in *italics*; references to tables are in **bold**.

Aberdeen Bestiary 66; *see also* bestiary, medieval
Accipiter sp. 27; *see also* hawk
Adelard of Bath 26, 28, 30, 32–3, 231; *see also* medieval
ADS *see* Archaeology Data Service
agency 2, 4, 114, 119, 147, 217–18, 229
agriculture 10, 13–14, 56, 71, 81, 85–6, 88, *91*–2, 94, 96–8, 127–31, 133–8, 142, 181–2, 189, 192–4; *see also* crops, farm, farming, livestock
Ailuropoda melanoleuca 179–81; *see also* giant panda
Alcedo sp. *see* kingfisher
Alces alces 153; *see also* elk
Anas sp. *see* duck
animal behaviour 191; *see also* boldness, charisma, fear, gregariousness, intelligence, nobility, tameness
animal diet *see* feeding, nutrition, *individual foods*
animal protection laws 64, 108, 207
Anser sp. *see* goose
anthropocentrism 2, 105, 123, 223, 226
anthropology 3–5, 19, 21, 26, 105, 181, 230, 232
Apis sp. *see* bee
Aquila sp. *see* eagle
archaeology 3–4, 10, 12, 19–21, 26, 29, 32, 34, 36, 39, 54–5, 62–3, 81, 83–4, 86, 88, *89*, 92, 97–8, 107, 109, 164, 191, 220, 225, 230–1
  bioarchaeology 3

zooarchaeology 62, 84, 86, 88, 98, 191, 220, 230
Archaeology Data Service 62, *89*
Ardea sp. *see* heron
arsenic 190, 194, 196–9; *see also* heavy metal, pollutants, toxic metal, poison
artefact 88, *90*, 91–2; *see also* Portable Antiquities Scheme
aurochs 132, 225; *see also Bos* sp., cattle, cow

badger **68**
bee 57, 223, 233–4
beef 127–8, 130, 134, 138–9, 166; *see also Bos* sp., cattle, cow
beetle 233
Belon, Pierre 65
bestiary 65, *66*, 93; *see also Aberdeen Bestiary*, medieval period
biodiversity 57, 108, 116, 126–9, 132–3, 135–7, 213, 220–2, 233–4
  functional biodiversity 57, 220, 222
biology 49–50, 54–5, 77, 127, 168, 172, 230; *see also* morphology
biopolitics 105, 114
bird 3, 5, 25–39, 46, 61–3, 65–7, 69–77, 81–8, 92, 94–9, 105, 107, 109–16, 118, 120, 126, 139–40, 158, 191, 195, 208–9, 211, 213–16, 222–5, 228–9, 231, 233–4; *see also* corvid, buzzard, duck, falcon, fowl, goose, harrier, hawk, pigeon, raptor, red kite, swan
bison 132, 154, 157–8

Boke of St Albans 27, 36, 38; *see also* medieval period
boldness 76, 191, 200, 209, 216, 229
bone 5, 33, 54–5, 63, 75, 86, 92, 127, 166, **167**–9, **170**, *171*, 172–3, 175, 178–9, 192–3, 195, 224–5, 231; *see also* archaeology, morphology, osteology, skeleton
Bos sp. *see* aurochs, cattle, cow
branch 88, 147–58, 175
Britain 12, 25–7, 29, 35–6, 52, 61–3, 67, 75, 77, 81, 83–4, 86, 88, *89*, *90*, 91–2, 94–6, 99, 123, 142, 189–92, 194–5, 205, 207–8, 210, 211, 213, 216–18, 222; *see also* UK
British Trust for Ornithology 96, 233
browse 147; *see also* branch, feeding: browse feeding
BTO *see* British Trust for Ornithology
Bubalus; see buffalo
buffalo 132
Buffon, Comte de 70
bullfinch **68**
butchery 64; *see also* animal products, animal services
Buteo buteo; see buzzard
butterfly 46, 223, 233–4; *see also* insect
buzzard 27, 37, 67, **68**, 71–2, 212, 222, 228; *see also* raptor

cadmium 190, 193, 196–9; *see also* heavy metal, pollutants, toxic metal
Callithrix sp. 175; *see also* marmoset
Canis familiaris; see dog
Canis lupus 124, 168; *see also* wolf
Capra sp. *see* goat
captivity 5, 25, 28, 30, 33–4, 126, 142, 148, 161–6, 168, 172, 175, **176**, 177–82, 205–**206**, 208–9, 223–5, 227, 232; *see also* park: wildlife park, zoo

care 4, 9–12, 14, 16–21, 25, 28–9, 31–2, 36, 48, 57, 64, 72–4, 77, 87, 106–9, 111–13, 115, 117–19, 124–7, 129, 134–5, 137, 142, 149, 154, 157, 161, 164–5, 177–9, 182; *see also* affective
cat 5, 19, 47, 50–1, 53, 55, 57, 71, 105–13, *111*, 119–120, 123, 126, 154, 166, 174–5, 234; *see also* Felis sp., wildcat
cat colony 106, 108–9, 111
cat lady 108–10, 112, 119; *see also* gattare
cat shelter 111
Cato the Elder 10–19, **15**; *see also* Roman
cattle 5, 9–21, **15**, 32, 123, 125, 127–8, 130, 132–5, 137–42, *141*, 155, 225–7, 231, 234; *see also* Aurochs, *Bos* sp., cow
Cavia porcellus 47; *see also* guinea pig
Cebuella sp. 175; *see also* marmoset
Cervus sp. *see* deer
charismatic animals 116, 207, 216, 233
chicken 46, 75, 125, 131, 133, 221; *see also* fowl
Chiltern Hills 76–7
chimpanzee 168, 174, 177–9, 182, 232; *see also Pan troglodytes*
chromium 190, 194–5, 196–9; *see also* heavy metal, pollutants, toxic metal
circus 5, 45, 47–52, 57, 126, 161, 210
Circus sp. *see* harrier
city 16, 48, 51, 54, 62–5, 67, 70–1, 73, 76, 82, 88, 105–7, 109, 112–15, 164, 191–2, 197–9, 207, 233; *see also* London, urban environment
climate change 127, 133, 223; *see also* sustainability
Coloeus sp. 81; *see also* corvid: jackdaw
Columba sp. 113; *see also* pigeon

Columella 10–12, 17–19; *see also* Roman
commensal 2, 4, 61–3, 65, 72, 77, 81, 97, 106, 109, 113, 115, 119, 191, 209, 224
companion animal *see* pet
conflict 5, 37, 61, 71, 74, 77, 81–5, 96–7, 105, 114, 128, **206**, 209, 217, 221, 227–8, 232
conservation 5, 49, 56, 61, 74–7, 135, 142, 152, 161–2, 164, 180, 182, 205–18, **206**, 231, 233
contamination 54, 72, 76, 114, 151, 189–90, 192–9; *see also* heavy metal, pollutants, toxic metal, waste
cormorant **68**, 70
corvid 5, 37, 63, 74, 77, 81–99, *89, 90*, **95**, 105, 117, 190–1, 212, 229; *see also Corvus* sp.
  chough **68**, 81, 94, **95**, 97; *see also Corvus* sp.
  crow **68**, 81–8, 91–**5**, 99, 107, 112–13, 116, 118; *see also Corvus* sp.
  jackdaw **68**, 81–2, 87, **95**; *see also Corvus* sp.
  jay 67, **68**, 81, 84–5, 93, **95**; *see also Corvus* sp.
  magpie 67, **68**, 81–2, 84–5, 87, **95** 116; *see also Corvus* sp.
  raven 67, **68**, 81–8, *91*–6, **95**, 98, 212, 221–2; *see also Corvus* sp.
  rook **68**, 81–2, 94–**5**; *see also Corvus* sp.
Corvus sp. 81; *see also* corvid
COVID-19 211, 214
cow 32, 48, 57, 126–7, 137–41, 226; *see also* aurochs, *Bos* sp., cattle
crane 32, 35; *see also Grus grus*
crop (bird anatomy) 31, 214; *see also* bird
crop (plants) 57, 62, 82, 84, 86–7, 92, 128, 131–2, 136, 193; *see also* agriculture, farm, farming
Ctenocephalides sp. 55; *see also* flea

cultural capital 135–6, 142
culture 2–3, 25, 33–4, 38, 51, 63, 65, 67, 77–8, 83, 86, 88, 91–2, 94, 96–8, 106–7, 109, 114–15, 119, 123, 125, 127–9, 135–7, 142, 221
  biocultural heritage 107
  cultural coevolution 97, 106
  nature-culture binary 127
Cygnus sp. *see* swan

De Avibus 65; *see also* bestiary
deer 32, 35, 157, 173
  fallow deer 229
  red deer 154, 157
  reindeer 153–8, 116, 135
Dendrocopos sp. *see* woodpecker
disease 9–10, 32, 34, 47, 50, 53–4, 55–7, 63, 85, 94, 96, 98, 108, 127, 165, 225; *see also* health, plague
DNA 55–6
dog 19, 35, 47–8, 50–1, 53, 55, 57, 105, 118, 124
Domesday Book 29; *see also* medieval
domestic 2, 3, 17, 19, 32, 39, 45, 47, 53, 55, 57, 66, 70–2, 76–7, 85, 105–6, 113, 119, 132, 162, 194, 224–5, 227, 232; *see also* agriculture, farm, farming, livestock
duck 1, 32, 116, 118, 224
dunnart 178; *see also Sminthopsis* sp.

eagle 27, 37–8, **68**, 70–1, 222
  golden eagle 222
  white-tailed eagle **68**, 222
early modern period 26, 32, 36, 62, 65, 67, 77, 88, 147; *see also* post-medieval
ecological services 56–7, 94, 98–9, 136, 207; *see also* disease, health, sanitary services
ecology 3–4, 56–7, 62, 67, 74–5, 77, 98–9, 105–7, 109, 113–19, 130, 132, 148, 220, 233–4

INDEX **239**

effective 1–2, 5, 9, 125; *see also* utilitarian
eggs 1, 9, 13, **15**, 17, 46, 51, 55, 67, **68**, 82, 85–7, **95**–7, 125–6, 135, 208; *see also* bird
elk 153–8; *see also Alces alces*
emotion 1–2, 9–10, 12, 19–20, 98, 125–7, 130, 137, 139; *see also* feeding: affective, non-utilitarian
endangered 29, 72, 95, 124, 163; *see also* extinction
enlightenment 19, 69, 71–72
entertainment services 45; *see also* circus, flea circus, park: wildlife park, zoo
entomology 223
environment 1, 3, 47, 55, 61–2, 73, 77, 81–2, 85, 96–8, 105–6, 113–14, 125, 127–33, 135–7, 142, 147, 155, 158–9, 161–5, 168, 172, 175, 177–8, 182, 189–4, 197–9, **206**, 209, 213, 224, 226–7, 229, 232
Equus sp. *see* horse
Erinaceus europaeus; see hedgehog
ethnography 4, 114, 125, 128, 130, 135, 147, 219, 231
extinction 5, 55, 73, 95, 124, 163, 179, 181–2, 207, 211, 213, 217, 222; *see also* endangered, extirpation
extirpation 211, 216–17; *see also* endangered, extinction
eyass 28–9; *see also* falcon, hawk, raptor

Falco sp. 27; *see also* falcon
falcon 25, 27–39, 225, 228; *see also* falconry, hawk
  gyrfalcon 27, 29, 32, 35–7; *see also* falconry, hawk
  hobby 27; *see also* falconry, hawk
  kestrel 27, **68**; *see also* falconry, hawk
  lanner 27; *see also* falconry, hawk
  merlin 27, 36; *see also* falconry, hawk
  peregrine falcon 27, 228; *see also* falconry, hawk
falconry 5, 25–37, 39, 63, 67, 87, 225, 228; *see also* hunting
farm 4, 5, 19, 123–7, 130, 133–4, 136, 139, 142, 210–11; *see also* agriculture, farming, good farming, livestock
farmer 57, 62, 67, 74–5, 84–5, 96–8, 123–43, *141*, 225–6, 234; *see also* agriculture, farm, farming, good farming, livestock
farming 4–5, 99, 123–31, 133–9, 213, 225, 234; *see also* agriculture, farm, good farming, livestock
  ethical farming 4; *see also* agriculture, farm, good farming, livestock
  intensive farming 133, 136; *see also* agriculture, farm, good farming, livestock
  regenerative farming 123, 125–6, 128, 130–6, 139, 142–3, 168, 234; *see also* agriculture, farm, good farming, livestock
fear 124, 140
feeding
  affective feeding 1–3, 5, 9, 20, 78, 83, 98–9, 119, 125–30, 133, 137, 139, 191, 220
  attractant feeding 205, **206**, 223
  browse feeding 125, 147, 151–7, 166, 177; *see also* branch, wildlife park, zoo
  captive feeding 148, 205, **206**, 208–9; *see also* captive
  conservation feeding 5, 49, 74, 205, **206**, 211–12, 216, 233
  diversionary feeding 82, 87, 118, 205, **206**, 209, 217, 223
  enrichment feeding 33, 49, 174–5, 178, 180, 182
  farm feeding *see* agriculture, farm, farming, good farming, livestock

forage feeding 62, 77, 82, 111, 113, 115, 125, 132–5, 138, 156, 177, **206**, 212, 226
for control 2, 50, 66, 71, 77, 112, 118–19, 126–7, 209, 220, 223–4
for education 98, 161, **206**, 211–12
for entertainment 5, 34, 45, 49, 51, 57, 67, 98, 126, 161, 166, **206**, 210, 216
garden feeding 76, 190, 216
graze feeding 14, 75, 95, 125, 128, 131–6, 140, *141*, 151, 157, 226; *see also* grazing
hand feeding 1, 28, 30, 48, 63, 113, 123–5, 209, 217
instrumental feeding 107, 115
intentional feeding 1, 3, 9, 17, 75, 84, 87, 93, 106, 110, 114, 125–6, 162, 190–1, 199–200, 205, **206**, 215, 217, 223–4
necro feeding 205, **206**, 223
non-utilitarian feeding 1–5, 9–10, 18, 20–1, 25, 27, 83, 98, 125–6, 183, 191, 219–20
opportunistic feeding 63, 189, 192, 205, **206**, 208, 214, 216–18, 223, 228
research feeding 205, **206**
scavenge feeding 5, 27, 37, 61–5, 70, 72, 75, 77, 81–2, 84–6, 93–4, 96–9, 115, 189, 192, 207–8, 212, 217, 229
supplementary feeding 30, 75, 115, 119, 134, 138, 190, 205, **206**, 208–10, 212, 214–15, 217, 223
tourism feeding 75, 126, 152, 205, **206**, 208, 210–11, 215–17, 223, 232; *see also* tourism
transactional feeding 125–6, 220
unintentional feeding 106, 114, 162, 190, 199–200, 205, **206**, 208, 210, 214–17, 223
veterinary feeding 5, 13, 19, 231; *see also* veterinary

wildlife feeding 75, 205, **206**, 223; *see also* wildlife
zoo feeding 5, 49, 226; *see also* zoo
Felis sp. *see* cat, wildcat
feral 2, 77, 106–9, 111, 113–14, 118–19, 123, 191; *see also* stray animals
ferret **170**
flea 5, 45–58, 126, 161, 210
  cat flea 50–1, 55; *see also* *Ctenocephalides* sp.
  dog flea 50–1, 55; *see also* *Ctenocephalides* sp.
  human flea 47–51, 53, 55, 57; *see also Pulex* sp.
  jigger flea 55; *see also Tunga* sp.
  rabbit flea 52; *see also Spilopsyllus* sp.
  rat flea 50, 54; *see also Xenopsylla cheopis*
flea circus 5, 48; *see also* flea
food production 4, 75, 126–7; *see also* farming, hunting
fowl 30, 35, 38, 66, 70, 116; *see also* chicken, duck, goose
fox 5, 61, **68**, 73, 77, 124, 189–93, 195–200, *196*, 224, 232; *see also Vulpes vulpes*
Frederick II 27–33, 35–7, 39; *see also* medieval
From 'Feed the Birds' to 'Do Not Feed the Animals' 3, 4, 219–21, 229–30, 233
fur 46, **176**, 191

Gallus domesticus; *see* chicken
garbage *see* waste, rubbish
Garrulus sp. 81; *see also* corvid: jay
gattare 108–10, 112, 119; *see also* cat lady
generalist 61, 81, 115, 209, 217, 228
geography 105, 135, 163, 166, 180
giant panda 179–81; *see also Ailuropoda melanoleuca*
goat 1, 125, 135

INDEX

good farming 85, 123, 127–30, 133–6, 141–2
goose 1, 116, 118; *see also* fowl
grazing 14, 75, 95, 125, 128, 131–6, 140, *141*, 151, 157, 226; *see also* graze feeding
   extended grazing 128, 136
   mob grazing 128, 131, 134, 141
   rotational grazing 125, 131–3
Grus grus 32, 35–6; *see also* crane
guinea pig 47, 53
gull 77, 109–10, 116
Gulosus aristotelis; *see* shag

Haliaeetus sp. *see* eagle
hare 30
harrier 67, **68**
hawk 25–7, 29–30, 32, 35, 38, **68**, 70–1, 114; *see also* falcon, falconry
   goshawk 27, 29, 35, 228; *see also* falcon, falconry
   Harris hawk 114; *see also* falcon, falconry
   sparrowhawk 27, 29, 35, 228; *see also* falcon, falconry
health 1, 3, 11, 14, 16–17, 20, 28–9, 31–2, 34, 39, 49, 55, 63, 76, 82, 86, 95, 99, 105–9, 112–16, 123, 126–8, 132–3, 135–7, 151, 161, 165–6, **167**, 179, 181–2, 189–90, 192–4, 197–200, 211, 226–7, 231, 234; *see also* disease, pathology, plague
   public health 107–8, 113–14, 211
heavy metal 189–90, 194, *196–9*, 224; *see also* contamination, toxic metal, pollutants, waste
hedgehog **68**
Henry VIII 36, 94; *see also* medieval
herbicide *see* pesticide
heron 38, 67, **68**, 116
hierarchy 9, 26–8, 36, 39, 105–7, 112–13, 116, 118–19; *see also* culture
history 5, 10–12, 22, 25, 45, 47–8, 51–2, 54, 57, 61–2, 65, 77, 81, 83–4, 86, 114, 178, 181, 207

honey 33, 57, 87, 223; *see also* bee, insect
hornet 109; *see also* Vespa orientalis
horse 10–11, 48, 113, 154–5, 157–8, 166
Hugh of Fouilly 38, 65–6; *see also* medieval period
human–animal studies 4–5, 19
hunter-gatherer 182, 209; *see also* prehistory
hunting 25–7, 30–9, 48, 66, 70, 72, 77, 111, **176**, 182, 191, **206**, 209, 212, 214, 221, 228; *see also* falconry, prey

Iceland 29
industry 48, 127, 129, 189, 192–5, 197–8
injury *see* trauma
insect 46, 48, 50–3, 56–7, 95, 107, 109–10, 191, 223, 233; *see also* bee, butterfly, flea
Insh Marshes 149–152
intelligence 81, 98
interdisciplinarity 3, 52, 219, 229, 232; *see also* multidisciplinarity
International Union for the Conservation of Nature 161–2
interspecies 4, 83–4, 106; *see also* multispecies
Ireland 229
isotope 19–21, 55
it-narratives 147–8
IUCN *see* International Union for the Conservation of Nature

kingfisher **68**

lagomorph *see* hare, rabbit
lamb 82, 85, 97–8, 124, 126, 134; *see also* sheep
Laridae *see* gull
lead 190, 192–3, 195, 196–9; *see also* heavy metal, pollutants, toxic metal, leaded petrol
leaded petrol 192, 196; *see also* lead, industry, pollutants, heavy metal

leather 1, 30, 125–6; *see also* cattle, cow
leopard 71, 154
Lepidoptera; see butterfly
lion 49, *171*; *see also Panthera* sp.
livestock 9, 11–12, 19–21, 45, 48, 53, 55, 57, 82–4, 86, 95, 98, 123–9, 131–9, 141–2, 151, 225–6; *see also* agriculture, domestic, *Bos* sp., cattle, cow, farm, farming, lamb, sheep
lockdown *see* COVID-19
London 29, 48, 64, 76–7, 83, 94, 96, 98, 106–7, 113–20, 130, 190–1, 195–7, *196*, 232; *see also* city, urban
London Wildlife Protection 115
lure 30, 39, **206**; *see also* falconry, hunting
Lutra sp. *see* otter

Macaca sp. 136, **170**, 178; *see also* macaque
macaque 163, **170**, 178, 232; *see also Macaca* sp.
  Barbary macaque 232; *see also Macaca* sp.
  Japanese macaque 178; *see also Macaca* sp.
  rhesus macaque 163, **170**; *see also Macaca* sp.
marmoset 175
marten **68**
Martes sp. *see* marten
meat 1, 9, 12, 29–30, 32–3, 37, 66, 72, 74–5, 113, 125–8, 135, 166, 173, *174*, **176**, 180, 191–3, 195, 215, 217, 225, 228; *see also* beef, pork
medieval period (European) 5, 10, 11, 25–9, 31–7, 39, 62–7, 76, 91, 93–5, 99, 207, 228
Meles meles; see badger
Mico sp. 175; *see also* marmoset
Middle Ages *see* medieval period (European)
milk 1, 12, 57, 124–6, 133, 166, 226; *see also* cattle, cow

Milvus sp. 61, 205, 207; *see also* red kite
modern period 4–5, 10–11, 19–21, 28, 34, 45, 47, 53–4, **68**, 74, 76, 81, 83–4, 87, 91–2, 96, 98, 113, 129, 155, 161, 164, 177, 181, 191, 197, 199, 215, 231; *see also* early modern period
mole **68**
morphology 5, 34, 115, 148, 161–**70**, **167**, 172–3, 177–83, 220, 224–5, 229, 230, 232–3
  morphological plasticity 34, 162–3, 168–**70**, **172**–3, 177–82, 224, 232
mouse **68**, **167**
multidisciplinarity 219, 229; *see also* interdisciplinarity
multispecies 4, 147; *see also* interspecies
Mus musculus **167**; *see also* mouse
Mustela sp. *see* ferret, stoat, polecat, weasel

natural 3, 17, 19, 25, 27–8, 30, 32–4, 38, 49, 52, 65, 70–4, 98–9, 132, 135, 156–7, 161, 163–5, 173–4, 177, 189, 194–5, **206**, 208, 212–13, 216, 220, 224–7, 232
nest, nesting 28–29, 46, 73, 75, 82, 85, 95–97, 113, 228
niche 61–2, 64, 72–3, 76–7, 106–7, 109, 113, 115–17, 222
nobility 36–7, 70–1
Norse 83, 91–2, 97
nutrition 33, 49, 75, 115, 131, 138, 157, 159, 163, 165–9, **167**, 172, 175, 179–81, **206**, 209, 226–7

osprey **68**
osteology 88, 91; *see also* archaeology, bone, morphology, skeleton
otter **68**
Ovis aries; see lamb, sheep

palaeontology 164
Pan troglodytes 168, 174, 179; *see also* chimpanzee
pandemic *see* COVID-19
Pandion haliaetus; *see* osprey
Panthera sp. *see* leopard, lion, tiger
parasite 5, 45–7, 53–7, 165, 230; *see also* pest, vermin
park 32, 77, 107, 112–13, 115–18, *117*, 147, 149–53, 155, 157–8, 210
   deer park 32
   public park 77, 107, 112–13, 115–18, *117*
   wildlife park 147, 149–53, 155, 157–8, 210
PAS *see* Portable Antiquities Scheme
pathology 164, 166, 177–8; *see also* disease, morphology
persecution 36, 45, 61, 64, 67, 71, 73–5, 77, 82–6, 94–9, **95**, 191, 207–8, 211, 221, 231
pest 2, 5, 67, 69, 72, 74–5, 77–8, 82–4, 107, 119, 191, 195, 208, 221; *see also* parasite, Tudor vermin laws, vermin
pesticide 131, 134, 137, 194; *see also* farming
pet 2, 4, 47, 53, 56, 87–8, 92, 107–9, 119, 124, 127
pet food 4, 109–12, 192
Phalacrocoracidae sp. *see* cormorant
Pica sp. 81; *see also* corvid – magpie
Picus sp. *see* woodpecker
pig, 30, 47, 125, 131, 133, **167**–168, **170**, 173; *see also Sus domesticus*
pigeon 5, 30, 32, 61, 77, 82, 84, 105–7, 109–10, 113–20, 123, 126, 190–1, 221, 224; *see also Columba* sp.
plague 50, 54, 93; *see also* disease, health
Pliny the Elder 87–8; *see also* Roman
poison 71–73, 93, 199; *see also* necro feeding, pollutants, heavy metal, toxic metal
polecat **68**

pollutants 5, 224; *see also* contamination, heavy metal, toxic metal, waste
pork 30; *see also* pig, *Sus domesticus*
Portable Antiquities Scheme 88, *90*, *91*, 95; *see also* artefact
Portable X-ray fluoresence 196–6
post-medieval 84, 93, 95; *see also* Enlightenment
prehistory 62, 84, 215; *see also* hunter-gatherer
prey 25–6, 30, 32–3, 35, 37–9, 50, 56, 63, 66–7, 70–1, 73, 75, 166, *171*, 173–4, **176**, 180, 197, 213, 229
Procavia capensis **170**, 179; *see also* rock hyrax
products, animal *see* bone, eggs, fur, honey, leather, meat, milk
Pulex Irritans 47; *see* flea
Pulex sp. 47–8, 51–3, 55; *see also* flea
pXRF *see* portable X-ray fluoresence
Pyrrhocorax sp. 81; *see also* corvid – chough
Pyrrhula sp. *see* bullfinch

rabbit 46, 53, 67, 72, 74–5, 95, 150, 191; *see also* flea: rabbit flea
raptor 19, 25, 27–8, 35–6, 38, 61–2, 67, **68**, 70, 209, 211, 213, 228; *see also* falcon, hawk, red kite
rarity 36, 221
rat 61, **68**, 72, 83, 115–18, *117*, **167**, **170**, 228; *see also* flea: rat flea, *Rattus* sp.
Rattus sp. **167**, **170**; *see also* rat
Reading 76, 221, 227–8; *see also* city, urban
recombinance 106, 113, 118–20
red kite 5, 27, 37, 61–78, **66**, 83, 94, 105, 189, 191, 205, 207–18, 221–2, 227–8, 230, 233; *see also Milvus* sp.
red squirrel 124; *see also* squirrel, *Sciurus vulgaris*

reintroduction 74–5, 161–2, 207–10, 221–2; *see also* rewilding, conservation
relationality 106
rewilding 74, 217, 222
Richard III (play) 71; *see also* Shakespeare, William
Ritual 9, 12, 14–21, 34–5, 63, 86, 91, 123
roadkill 75, 85, 214; *see also* scavenge feeding
Rock hyrax **170**, 179, 181; *see also Procavia capensis*
Roman 5, 9–21, 27, 32, 47, 83, 86–8, *91*–2, 97, 108, 123, 226–7, 231
Rome 12–14, 16, 87, 106–7, *111*, 112–13, 119–20
Royal Society for the Protection of Birds, the *see* RSPB
RSPB 74
rubbish 63, 65, 85–6, 99, 192
rural 67, 76–7, 82, 107, 191, 198, 222

sanctuary 4, 124, 168
sanitary services; *see also* disease, health, ecological services
scapegoat 85, 97
scavenging *see* feeding: scavenge feeding 62–5, 72, 77, 96, 115, 189, 192, 207–8, 229
Schaseck of Bohemia 64
Sciurus carolinensis 56; *see also* red squirrel, squirrel
Sciurus vulgaris 56, 124; *see also* red squirrel, squirrel
services, animal *see* ecological services, sanitary services, entertainment services, traction
shag **68**
Shakespeare, William 70; *see also Winter's Tale, The, Richard III (play)*
sheep 57, 75, 85, **95**, 97, 125, 128, 134–5, 138–9, 149, 151, 234; *see also* lamb
Siphonaptera sp. 46; *see also* flea

skeleton 33, 36, 54–5, 62–3, 86, 163–9, **167**, **172**–8, 180, 182, 193, 225; *see also* bone
skilled vision 137–8, 141
Sminthopsis sp. 178; *see also* dunnart
socioculture *see* culture
sociology 135, 220
soil health 128–9, 131, 132, 226
Spain 74, 208
Spilopsyllus sp. 53; *see also* flea: rabbit flea
sport 11, 25–7, 29, 31, 34, 73, 110, 191; *see also* falconry, hunting
squirrel 124; *see also* red squirrel, *Sciurus vulgaris*
starling **68**
stoat **68**
stray (animals) 2, 5, 105, 107–8, *111*, 113, 123, 126, 234; *see also* feral
stringfoot 115, 117
strychnine *see* poison
Sturnus sp. *see* starling
Sus domesticus **167**, **170**; *see also* pig
sustainability 127, 131, 136, 158
swan 67, 224
Sweden 74, 92, 99, 208; *see also* Norse

Talpa europaea see mole
tameness 2–3, 30–1, 35, 38–9, 63, 77, 87–8, **206**, 209, 228
Taylor, John 65
tiger 49, 57, 163, 173, *174*, **176**, 179–80; *see also Panthera* sp.
tourism 52, 75, 98, 126, 152, 205, **206**, 208, 210–11, 215–17, 223, 232; *see also* feeding: tourism feeding
tourist *see* tourism
toxic metal 190, 192–3, 199; *see also* contamination, heavy metal, pollutants, waste
traction 1, 12, 125–6; *see also* cattle, cow, horse
trap 72, 108
trash *see* rubbish

trauma 34, 115, 140, 177; *see also* disease, health, pathology
trophic order/levels 36–7, 56
Tudor vermin laws 66–7, **68**, 71–2, 94–6, 208; *see also* pest, vermin
Tunga sp. 55; *see also* flea
Turner, William (naturalist) 70

United Kingdom 49, 56, 61–2, 75–6, 78, 82, 84, 86, 95, 98, 124, 132, 134, 191–2, 211, 213, 216; *see also* Britain, London, Reading
urban environment 1, 5, 29, 37, 61–5, 67, 76–7, 82–4, 94, 96, 105–9, 112–15, 118–20, 189–94, 197–9; *see also* city, London
utilitarian 1–5, 9–10, 12, 18, 20–1, 25, 27, 78, 83, 94, 98, 119, 125–6, 137, 139, 183, 191, 219–20, 222; *see also* effective

vermin 48, 65–7, **68**, 69–74, 94–6, 208, 220; *see also* parasite, pest, Tudor vermin laws
vermin Acts *see* Tudor vermin laws
Vespa orientalis 109; *see also* hornet
veterinary 5, 9–11, 13–14, 16–21, 32, 107–8, 123–4, 127, 164–5, 177–8, 195, 214, 231; *see also* feeding: veterinary feeding
Vulpes vulpes 190; *see also* fox

Wales 74, 76, 207, 210–11
waste 61–5, 70, 75–6, 84, 86, 94, 96–7, 99, 114, 139, 149–50, 157–8, 165, 189–91, 194–5, 197–200, **206**; *see also* contamination, rubbish, heavy metal, pollutants, toxic metal
Waterman Poet *see* Taylor, John

weasel **68**
wild 2–3, 5, 26, 30–3, 35–6, 38–9, 49, 66, 73–4, 76–8, 82–4, 87–8, 92–3, 95, 99, 111, 116, 119, 123–4, 126, 132, 138, 154, 161–6, 168, 172, 175–82, **176**, 190, 193, 205, 209, 212, 224–8, 232
wildlife 2, 64, 67, 73–4, 77, 107, 115–16, 118, 124, 142, 165, 190–1, 198, 205, **206**, 207, 210, 216–18, 223, 233–4
wildness 66, 72, 74, 77–8
wildcat **68**, 150, 154; *see also* cat, *Felis* sp.
Wildlife and Countryside Act, 1981 64
wildlife gardening 223–4
wine 13, **15**, 17, 30
Winter's Tale, The 70; *see also* Shakespeare, William
wolf 124, 154, 226; *see also Canis lupus*
woodpecker **68**
wool 124, 126, 135; *see also* lamb, sheep

Xenopsylla cheopis 50, 54; *see also* rat flea

Yersinia pestis 50, 54; *see also* rat flea, *Xenopsylla cheopis*

zoo 1, 5, 49, 124, 126, 148, 153, 155, 157, 161–6, 168, *171*, 173, *174*–5, 177, 179, 180, 182, 210, 226–7, 230; *see also* nutrition
zoology 3, 52, 147, 164, 168; *see also* archaeology: zooarchaeology, biology